○●○○○

Mathematics
in Daily Living
REVISED

FRACTIONS

Nerissa Bell Bryant
Staff Consultant
Adult Performance Level Project
Louisiana Tech University

Loy Hedgepeth
Director of Adult Education
Ouachita Parish Schools
Monroe, Louisiana

Steck-Vaughn Company ○ Austin, Texas

Introduction

This book contains adult-oriented instructional material designed to teach mathematics skills and life-coping skills to the mature learner.

The academic skill content of this book was determined by data compiled from a survey of adult education teachers. This survey revealed the topics for which teachers most urgently needed teaching materials.

In addition to the mathematics skill content, this book focuses on life-coping skills needed daily by adults. The curriculum emphasis focuses on these five general areas: health, government and law, consumer economics, community resources, and occupational knowledge. Information on various topics in these five areas is presented along with practice of mathematics skills. Thus, as learners progress through the mathematics units, they are afforded an opportunity to (1) develop academic skills related to *mathematics competence* and (2) develop life-coping skills related to *functional competence*.

During their development these materials were field-tested at the Northeast Louisiana Learning Center in Monroe, Louisiana.

How To Use This Book

Each unit contains individualized instruction for self-development. However, the format and use of the book should be understood in order to expedite progress and maximize proper use of the material.

Unit: A section providing instruction, examples, practices, reviews, and evaluations of a designated skill. Units are subdivided into lessons.

Skill: Each intermediate skill necessary to the mastery of the designated unit skill is presented in each lesson title.

Instruction: Explanation and/or definition of each skill is offered. Examples follow each segment of instruction to clarify the instruction and prepare the learner to undertake practice exercises.

Exercises: Exercises to be worked and checked by the learner. Exercise B provides the same type of practice as Exercise A.

Review: An exercise providing practice and review on skills taught in the unit.

Coping Skills: Most units end with an activity designed to develop greater competency in life-coping skills. Other activities related to coping skills are within the lessons.

Answers: Answers to all pretests, exercises, and reviews are provided at the back of the book.

NOTICE: Answer Key is bound in the back of the book.

ISBN 0-8114-1513-9

7 8 9 0 C 94 93 92 91 90

Contents

UNIT 1—PRETEST
REVIEWING WHOLE NUMBERS

Write the place value name for the *4* in each number.

1. 724,306 _____

2. 343,105,662 _____

3. 459,823 _____

4. 47 _____

Add the following numbers.

5.
```
    8 8 4
    3 2 3
  1,6 0 7
+   9 9 7
```

6. 60,008 + 234,509 + 879,633 = _____

7.
```
   2,4 5 9
   3,7 8 1
     1 5 2
   9,7 3 4
+ 5 0,8 8 9
```

8. 953,069 + 104,663,911 + 23 + 7 = _____

Subtract the following numbers.

9.
```
  7,2 4 5
- 3,0 1 2
```

10. 82,896 – 968 = _____

11.
```
  4,0 0 0
- 3,6 4 7
```

12. 20,001 – 10,736 = _____

Multiply the following numbers.

13.
```
  5,0 7 3
× 3 5 2
```

14. 199 × 9 = _____

15.
```
  9 5,4 2 1
×     7 4 3
```

16. 74,742 × 3 = _____

Divide the following.

17. $2\overline{)794}$

18. $45\overline{)92,335}$

19. $37\overline{)10,184}$

20. $132\overline{)68,233}$

Read and solve these problems. Blacken the letter to the right that corresponds to the correct answer.

21. Carlene has a 436-page book to read for one of her night classes. She read 50 pages a day for 5 days but now has only 2 days left to complete the book. How many pages must she read each day?
a. 218   b. 93   c. 96   d. 154

[a] [b] [c] [d]

22. When Teresa and her two roommates moved into their new apartment, they had to put up a deposit of $300, pay the first month's rent of $550, and send the moving company a payment of $182 for transporting their furniture. How much was each person's share of the expenses?
a. $344   b. $324   c. $244   d. $216

[a] [b] [c] [d]

23. The Cabiyas made furniture payments of $36 a month for 35 months and a final payment of $35. How much did they pay for the furniture?
a. $2,195   b. $2,915   c. $1,295   d. $1,395

[a] [b] [c] [d]

24. When they first married, Mark and Susan bought a house with 1,600 square feet of living space. Last year they sold the house and bought another house which is 437 square feet larger. What is the size of Mark and Susan's new house?
a. 2,037 sq. ft.   b. 2,007 sq. ft.   c. 1,037 sq. ft.
d 1,637 sq. ft.

[a] [b] [c] [d]

25. The owners of Azzari Farms were drawing plans for planting a pecan orchard of 100 trees. They wanted to have the same number of rows as there would be trees in each row. How many rows of trees will there be?
a. 8   b. 9   c. 12   d. 10

[a] [b] [c] [d]

26. How many trees will be in each row in the new Azzari orchard?
a. 10   b. 12   c. 8   d. 9

[a] [b] [c] [d]

2

REVIEWING WHOLE NUMBERS

Naming the Periods and Place Values of Positions in Numbers

Instruction

The value of a digit depends upon its place in a number. The chart below shows the place value of 9 positions.

Millions' Period			Thousands' Period			Ones' Period		
Hundred Millions	Ten Millions	Millions	Hundred Thousands	Ten Thousands	Thousands	Hundreds	Tens	Ones
8	4	3 ,	6	2	4 ,	5	1	8

Each group of three digits, starting from the right, is called a period. Notice that the periods are separated by commas.

Example

6 in 651,832,014 indicates **hundred millions.**

4 in 58,142 indicates **tens.**

1 in 421,075 indicates **thousands.**

Exercise A

Write the place value name for each underlined digit.

1. 3<u>2</u>4,717 ....

2. 14,62<u>5</u> ....

3. 61<u>7</u>,258,129 ....

4. 504,2<u>3</u>1,644 ....

5. <u>6</u>2 ....

6. 91<u>3</u> ....

7. 1,25<u>2</u>,646 ....

8. 8<u>5</u>,421,560 ....

9. <u>9</u> ....

10. 12<u>7</u> ....

Exercise B

Write the place value name for the 6 in each number.

1. 61 ....

2. 214,623 ....

3. 1,621 ....

4. 46 ....

5. 6 ....

6. 4,768 ....

LESSON TWO: Adding Whole Numbers

Instruction

The addition of whole numbers is the combining of single- or multiple-digit numbers to find their sum.

Example

Single-Digit Numbers:	9	Addend
	+ 9	Addend
	1 8	Sum

	1 1	
Multiple-Digit Numbers:	4 1 9,6 5 0	Addend
	+ 2 8,6 0 5	Addend
	4 4 8,2 5 5	Sum

Exercise A   Add the following numbers.

1. 6
    + 5

2. 0
    + 9

3. 8
    + 4

4. 3 8
    + 5 1

5. 2 6 4
    + 3 2 3

6. 7 6
    2 4
    + 8 7

7. 4 5
    1 5
    + 7 9

8. 4 8 8
    6 0 9
    + 3 5 9

9. 4,7 6 2
    9,3 7 4
    1,2 9 8
    + 3 0 7

10. 5,6 3 5
    3 6 8
    1,2 7 5
    + 8 1 5

11. 3,000 + 12 + 4 + 190 =

12. 1,232 + 943 + 28 + 609 =

13. 11,426 + 385 + 17,201 + 100,371 =

14. 80,001 + 14,992 + 6,843 + 7 =

15. 34 + 7 + 426 + 3,110 + 71,436 =

16. 200 + 5,000 + 70 + 2 =

17. 500 + 2,606 + 17 + 60,582 =

18. 1,001 + 14 + 11,400 + 983 + 144 =

19. 138 + 228 + 17 + 432,678 =

4

20. 1,134 + 99 + 4,009 + 16 + 147 =

21. 5,365 + 368,323 + 264 + 518 =

22. 7,624 + 154,579 + 5,000 + 307 =

23. 49 + 62,411 + 411 + 14,440 + 9 =

24. 359,609 + 79 + 4,726 + 83 =

Exercise B   Add the following numbers.

1.	2.	3.	4.	5.
9 + 9	26 + 32	64 + 23	47 + 45	4,762 + 307

6.	7.
283 841 + 473	530 727 + 943

8. 21,489 + 9 + 5 + 0 =

9. 20 + 5 + 10 + 1 =

Use addition to complete the card below.

INVENTORY CARD

Description	Model No.	Date: 7/31 Qty.	8/31 Qty.	9/30 Qty.	10/31 Qty.
VHS Recorder	273A	9	7	15	11
VHS Recorder	473	3	12	12	8
VHS Recorder	976	4	14	7	3
VHS Recorder	P97	13	10	9	4
Sub Total					
Beta Recorder	327	2	12	12	10
Beta Recorder	473	5	10	8	8
Beta Recorder	B806	2	2	2	1
Sub Total					

10. How many VHS recorders were in stock on August 31? ----------------

11. On October 31, how many recorders of both types were in stock? ----------------

5

LESSON THREE: Subtracting Whole Numbers

Instruction   When subtracting whole numbers, you find the difference between a large number and a smaller number.

Example

Single-Digit Numbers:
$$5 \leftarrow \text{Minuend}$$
$$- 3 \leftarrow \text{Subtrahend}$$
$$2 \leftarrow \text{Difference}$$

Multiple-Digit Numbers:
$$\overset{3\ 12}{4,279} \leftarrow \text{Minuend}$$
$$- 1,363 \leftarrow \text{Subtrahend}$$
$$2,916 \leftarrow \text{Difference}$$

Exercise A   Subtract the following numbers.

1. 9 − 8	2. 7 − 3	3. 14 − 6	4. 16 − 7	5. 389 − 252
6. 71 − 27	7. 2,370 − 1,890	8. 8,507 − 2,939	9. 1,725 − 145	10. 2,000 − 1,425

11. 5,023 − 3,247 =   12. 2,983 − 898 =   13. 896 − 689 =

14. 2,321 − 673	15. 655 − 493	16. 8,011 − 7,322	17. 86,892 − 9,254	18. 420 − 89

Exercise B   Find the differences.

1. 13 − 7	2. 15 − 8	3. 14 − 13	4. 896 − 534	5. 722 − 435

6. 5,000 − 393 =   7. 968 − 732 =   8. 46,689 − 689 =

9. 8,239 − 989 =   10. 2,370 − 1,089 =   11. 5,271 − 415 =

6

Adding and Subtracting Whole Numbers

Directions   Read the graph. Use information from the graph to answer the questions which follow.

ELECTRIC SHAVER
Sales Volume
(In Hundreds)

	1	2	3	4	5	6	7	8	9
January	▓	▓	▓						
February	▓	▓	▓	▓					
March	▓	▓	▓	▓					
April	▓	▓	▓	▓					
May	▓	▓	▓	▓	▓	▓	▓		
June	▓	▓							
July	▓	▓	▓						
August	▓	▓	▓	▓					
September	▓	▓	▓	▓	▓	▓			
October	▓	▓	▓	▓	▓	▓			
November	▓	▓	▓	▓	▓	▓	▓		
December	▓	▓	▓	▓	▓	▓	▓	▓	

1. The highest shaver sales occurred in December, when 840 were sold. How many fewer were sold in November?   ....

2. Find the month when the fewest shavers were sold. How many more shavers were sold during the following month?   ....

3. How many shavers were sold from August through November?   ....

7

Multiplying Whole Numbers

Instruction Multiplication of whole numbers is a shortcut for addition.

Example

> 4×8 is the same as having four 8s added together.
> $8 + 8 + 8 + 8 = 32$
> $4 \times 8 = 32$

Multiplicand	3 2 7	**First:** Multiply 2×327. Write 654.
Multiplier	× 4 2	
Partial Products	6 5 4	**Next:** Multiply 4×327. Write 1308. Remember to in-
	1 3 0 8	dent one space.
Product	1 3,7 3 4	**Next:** Add the partial products to find the product.

NOTE: You indent one space each time you multiply by a different digit in the multiplier.

Exercise A Multiply the following numbers.

1. 7
   × 4

2. 8
   × 9

3. 7
   × 8

4. 6
   × 9

5. 7 1
   × 7

6. 9 9
   × 7

7. 5 8
   × 6

8. 2 3 3
   × 3 5

9. 4 0 0
   × 3

10. 5,0 2 3
    × 8 0 7

11. $243 \times 2 =$

12. $496 \times 6 =$

13. $541 \times 3 =$

Exercise B Find the products.

1. 7
   × 9

2. 9
   × 9

3. 6
   × 8

4. 5
   × 9

5. 6 9
   × 8

6. $81 \times 9 =$

7. $72 \times 8 =$

8. $3,333 \times 9 =$

Dividing Whole Numbers

Instruction

The division of whole numbers involves determining how many units of the divisor are contained in the dividend.

Example

```
 ┌Divisor
 │ Dividend─┐
 │ Quotient─┐
 │        136◄─┐
 └►24)3,264◄───┘
       -24
        86
       -72
        144
       -144
```

First: $32 \div 24 = 1$. Write 1 in quotient above 2. $1 \times 24 = 24$. Write 24 below 32.

Next: $32 - 24 = 8$. Write 8 under 4. Bring down 6.

Next: $86 \div 24 = 3$. Write 3 in quotient above 6. $3 \times 24 = 72$. Write 72 below 86.

Next: $86 - 72 = 14$. Write 14 under 72. Bring down 4.

Next: $144 \div 24 = 6$. Write 6 in quotient above 4. $6 \times 24 = 144$. Write 144 below 144.

Next: $144 - 144 = 0$. The 0 usually is not written.

There are 136 units of 24 in 3,264.

NOTE: Not all problems will contain an even number of units. Some will have a remainder. The remainder must be smaller than the divisor.

```
      8 r 1
   4)33
    -32
      1
```

Exercise A

Divide the following numbers.

1. $6)\overline{24}$ 2. $7)\overline{70}$ 3. $9)\overline{45}$ 4. $2)\overline{184}$ 5. $6)\overline{496}$

6. $4)\overline{248}$ 7. $42)\overline{8,610}$ 8. $34)\overline{3,265}$ 9. $3)\overline{17,120}$

10. $400 \overline{)8,000}$

11. $526 \overline{)11,048}$

12. $427 \overline{)10,248}$

13. $5 \overline{)8,166}$

14. $22 \overline{)4,410}$

15. $4 \overline{)35,211}$

16. $3 \overline{)39}$

17. $5 \overline{)1,968}$

18. $9 \overline{)18,036}$

Exercise B   Divide.

1. $8 \overline{)48}$

2. $7 \overline{)35}$

3. $5 \overline{)535}$

4. $3 \overline{)504}$

5. $2 \overline{)681}$

6. $73 \overline{)24,747}$

7. $282 \overline{)65,706}$

8. $989 \overline{)27,692}$

9. $23 \overline{)13,064}$

10. $36 \overline{)28,623}$

11. $61 \overline{)15,768}$

12. The Reyna's have 36 plants to distribute to 6 nursing homes. If they give an equal number of plants to each home, how many plants will each place receive? -----------------

13. As a company representative, Jane has traveled 319,205 miles in the last 5 years. If she traveled approximately the same number of miles each year, what was her annual mileage? -----------------

10

STUDYING AN INCOME TAX FORM

Who must file an income tax return? The federal government may amend income tax regulations from year to year. Therefore, the requirements for determining who must file a return, amounts of exclusions and credits, and tax rates frequently change.

The Internal Revenue Service (IRS) provides free booklets which contain each year's regulations. These booklets are generally mailed to taxpayers. Many banks and post offices also have the booklets. The IRS also has toll-free phone numbers that you can call to have questions answered.

This chart shows who was required to file a federal income tax return for 1983.

You must file a tax return if— Your marital status at the end of 1983 was:	and your filing status is:	and at the end of 1983 you were:	and your gross income was at least:
Single (including divorced and legally separated)	Single or Head of household	under 65 65 or over	$3,300 $4,300
Married with a dependent child and living apart from your spouse all year	Single or Head of household	under 65 65 or over	$3,300 $4,300
Married and living with your spouse at end of 1983 (or on the date your spouse died)	Married, joint return	under 65 (both spouses) 65 or over (one spouse) 65 or over (both spouses)	$5,400 $6,400 $7,400
	Married, separate return	any age	$1,000
Married, not living with spouse at end of 1983	Married, joint return	any age	$1,000
	Married, separate return	any age	$1,000
Widowed in 1982 or 1981 and not remarried in 1983	Single or Head of household	under 65 65 or over	$3,300 $4,300
	Qualifying widow(er) with dependent child	under 65 65 or over	$4,400 $5,400
Widowed before 1981 and not remarried in 1983	Single or Head of household	under 65 65 or over	$3,300 $4,300

Other filing requirements—

You must also file a tax return if any one of the following applied for 1983:

either you could be claimed as a dependent on your parents' return and you had $1,000 or more in gross income that was not earned income—for example, taxable interest and dividends. (You must file Form 1040.)

or you owe any special taxes, such as social security tax on tips you did not report to your employer. (You must file Form 1040.)

or you owe uncollected social security tax or RRTA tax on tips you reported to your employer. (You must file Form 1040.)

or you received any advance earned income credit (EIC) payments from your employer.

or you earned $400 or more from self-employment after you deduct business expenses. (You must file Form 1040.)

or you were allowed to exclude income from sources within U.S. possessions and had gross income of $1,000 or more. (You must file Form 1040.)

The three forms used for individual tax returns are the 1040EZ, the 1040A, and the 1040.

Directions Look at the following 1040EZ instructions and form. Assume that you are John Brown and complete the form using the following information. Some items on the form have been completed for you.

John T. Brown is unmarried and has no children. He lives at 3700 Millway in Hometown, Maryland. His zip code is 01234. John's social security number is 516-04-1492. John, a clerk, wants to file a tax return on form 1040EZ. He is neither blind nor over 65 years of age. He wants to contribute to the presidential election campaign fund. His income was $19,450, and he received $125 in interest on a savings account. He received no dividends. The total amount of withholding deducted from his pay was $3,450.

1983 **Instructions for Form 1040EZ**

You can use this form if:

- Your filing status is single.
- You do not claim exemptions for being 65 or over, OR for being blind.
- You do not claim any dependents.
- Your taxable income is less than $50,000.
- You had **only** wages, salaries, and tips, and you had interest income (other than All-Savers interest) of $400 or less.

If you can't use this form, you must use Form 1040A or 1040 instead. See pages 4 through 6. If you are uncertain about your filing status, dependents, or exemptions, read the step-by-step instructions for Form 1040A that begin on page 6.

Completing your return

It will make it easier for us to process your return if you do the following:

1. Keep your numbers inside the boxes.
2. Try to make your numbers look like these `1234567890`
3. Do not use dollar signs.

Name and address

Use the mailing label we sent you. After you complete your return, carefully place the label in the name and address area. Correct any errors right on the label. If you don't have a label, print the information on the name and address lines.

Presidential election campaign fund

Congress set up this fund to help pay for Presidential election campaigns. You may have one of your tax dollars go to this fund by checking the box.

Figure your tax

Line 1. Write on line 1 the total amount you received in wages, salaries, and tips. This should be shown on your 1983 wage statement, **Form W-2**, (Box 10). If you don't receive your W-2 form by February 15, contact your local IRS office. Attach the first copy or Copy B of your W-2 form(s) to your return.

Line 2. Write on line 2 the total interest income you received from all sources, such as banks, savings and loans, and credit unions. You should receive an interest statement (usually **Form 1099-INT**) from each institution that paid you interest.

You cannot use Form 1040EZ if your total interest income is over $400, or you received interest income from an All-Savers Certificate.

Line 4. You can deduct 25% of what you gave to qualified charitable organizations in 1983. But if you gave $100 or more, you can't deduct more than $25. Complete the worksheet on page 19 to figure your deduction, and write the amount on line 4.

Line 6. Every taxpayer is entitled to one $1,000 personal exemption. If you are also entitled to additional exemptions for being 65 or over, for blindness, for your spouse, or for your dependent children or other dependents, you cannot use this form.

Line 8. Write the amount of Federal income tax withheld. This should be shown on your 1983 W-2 form(s) (Box 9). If you had two or more employers and had total wages of over $35,700, see page 23. If you want IRS to figure your tax for you, complete lines 1 through 8, sign, and date your return. If you want to figure your own tax, continue with these instructions.

Line 9. Use the amount on line 7 to find your tax in the tax table on pages 29-34. Be sure to use the column in the tax table for **single** taxpayers. Write the amount of tax on line 9. If your tax from the tax table is zero, write 0.

Refund or amount you owe—Compare line 8 with line 9.

Line 10. If line 8 is larger than line 9, you are entitled to a refund. Subtract line 9 from line 8, and write the result on line 10.

Line 11. If line 9 is larger than line 8, you owe more tax. Subtract line 8 from line 9, and write the result on line 11. Attach your check or money order for the full amount. Write your social security number and "1983 Form 1040EZ" on your payment.

Sign your return

You must sign and date your return. If you pay someone to prepare your return, that person must also sign it below the space for your signature and supply the other information required by IRS. See page 25.

Mailing your return

File your return by **April 16, 1984**. Mail it to us in the addressed envelope that came with the instruction booklet. If you don't have an addressed envelope, see page 28 for the address.

USGPO   1983 423 108   36-2249473

Department of the Treasury · Internal Revenue Service

1983

Form 1040EZ Income Tax Return for Single filers with no dependents (0)

OMB No. 1545-0675

Name & address

If you don't have a label, please print:

Please write your numbers like this.

1 2 3 4 5 6 7 8 9 0

Write your name above (first, initial, last)

Social security number

Present home address (number and street)

City, town, or post office, state, and ZIP code

Presidential Election Campaign Fund
Check box if you want $1 of your tax to go to this fund. ▶

Dollars   Cents

Figure your tax

1 Wages, salaries, and tips. Attach your W-2 form(s).   1

2 Interest income of $400 or less. If more than $400, you cannot use Form 1040EZ.   2

Attach Copy B of Form(s) W-2 here

3 Add line 1 and line 2. This is your **adjusted gross income.**   3

4 Allowable part of your charitable contributions. Complete the worksheet on page 19. Do not write more than $25.   4

5 Subtract line 4 from line 3.   5

6 Amount of your personal exemption.   6 | 1 | 000 | 00 |

7 Subtract line 6 from line 5. This is your **taxable income.**   7

8 Enter your Federal income tax withheld. This should be shown in Box 9 of your W-2 form(s).   8

9 Use the tax table on pages 29-34 to find the **tax** on your taxable income on line 7. Write the amount of tax.   9 | 3 | 236 | 00 |

Refund or amount you owe

10 If line 8 is larger than line 9, subtract line 9 from line 8. Enter the **amount of your refund.**   10

Attach tax payment here

11 If line 9 is larger than line 8, subtract line 8 from line 9. Enter the **amount you owe.** Attach check or money order for the full amount, payable to "Internal Revenue Service."   11

Sign your return

I have read this return. Under penalties of perjury, I declare that to the best of my knowledge and belief, the return is true, correct, and complete.

Your signature   Date

X

For IRS Use Only—Please do not write in boxes below.

For **Privacy Act and Paperwork Reduction Act Notice, see page 38.**

13

UNIT 2—PRETEST
ROMAN NUMERALS

Write roman numerals for these arabic numerals.

1. 5 _____   2. 3 _____   3. 2 _____   4. 4 _____

Write arabic numerals for these roman numerals.

5. VIII _____   6. VI _____   7. X _____   8. IX _____

Write roman numerals for these arabic numerals.

9. 11 _____   10. 29 _____   11. 15 _____   12. 18 _____

Write arabic numerals for these roman numerals.

13. L _____   14. LXI _____   15. LXXIX _____   16. XLVI _____

Write roman numerals for the arabic numerals below.

17. 1,600 _____   18. 98 _____

19. 450 _____   20. 613 _____

Read and solve these problems. Blacken the letter to the right that corresponds to the correct answer.

21. The hour hand on the bank clock is pointing to IX. What time is it?
    a. 11 o'clock   b. 9 o'clock   c. 4 o'clock   d. 6 o'clock
    ☐a ☐b ☐c ☐d

22. Mr. Rivera's English class has studied chapters I-XXXVI in the textbook. The last chapter in the book is XLII. How many chapters are left to be studied? The answer is in roman numerals.
    a. VI   b. XII   c. VIII   d. IV
    ☐a ☐b ☐c ☐d

23. When Marcela entered the Federal Building, she glanced at the clock by the elevator. The hour hand was on XII and the minute hand was on V. When she left, the hour hand was on II and the minute hand on V. How long was Marcela in the building?
    a. 3 hours   b. 4 hours   c. 1½ hours   d. 2 hours
    ☐a ☐b ☐c ☐d

24. The copyright date at the end of an episode of Perry Mason was MCMLXI. Write the copyright date in arabic numerals.
    a. 1971   b. 1961   c. 1951   d. 1981
    ☐a ☐b ☐c ☐d

ROMAN NUMERALS

Writing Roman Numerals 1–5

Instruction

Roman numerals, although they may seem strange, are not difficult to learn. Roman numerals are sometimes used to indicate chapters in books, paragraphs in outlines, and copyright dates. The numbers we usually use are called **arabic numerals**.

Example

Arabic	Roman
1	I
2	II
3	III

The roman numeral for 5 looks like a capital *V*. 5 = V. In roman numerals, when a smaller number (like I) comes in front of a larger number (like V), you subtract the smaller number from the larger number. The roman numeral for 4 is IV (5 − 1 = 4).

Example

Arabic	Roman
4	IV
5	V

Exercise A

Write roman numerals for the arabic numerals and arabic numerals for the roman numerals.

1. 2 _____   2. 3 _____   3. 1 _____

4. 5 _____   5. 4 _____   6. III _____

7. I _____   8. V _____   9. II _____

10. IV _____   11. 3 _____   12. 2 _____

Exercise B

Write roman numerals for the arabic numerals and arabic numerals for the roman numerals.

1. 4 _____   2. 1 _____   7. V _____

3. 3 _____   4. 5 _____   8. III _____

5. 2 _____   6. IV _____   9. I _____

Writing Roman Numerals 5 to 10

Instruction

To progress from 5 to 8 in roman numerals, you just add the roman numerals you already know to V.

Example

5 = V
6 = V + I or VI
7 = V + II or VII
8 = V + III or VIII

The Romans used X to mean 10. To write 9, simply put I in front of X. 9 = IX (10 − 1 = 9).

Example

9 = IX
10 = X

Exercise A

Write arabic numerals for the roman numerals and roman numerals for the arabic numerals.

1. 5 _____
2. 8 _____
3. 10 _____
4. 1 _____
5. 2 _____
6. 7 _____
7. 3 _____
8. 9 _____
9. 4 _____
10. 6 _____
11. VIII _____
12. X _____
13. V _____
14. I _____
15. II _____
16. VII _____
17. III _____
18. IX _____

Exercise B

Write arabic numerals for the roman numerals and roman numerals for the arabic numerals.

1. VIII _____
2. 8 _____
3. VI _____
4. V _____
5. 1 _____
6. 2 _____
7. II _____
8. 7 _____
9. X _____
10. III _____
11. 9 _____
12. 10 _____
13. IV _____
14. 6 _____
15. III _____
16. X _____
17. 5 _____
18. 4 _____

LESSON THREE: Writing Roman Numerals 10 to 29

Instruction

To count from 10 to 19 in roman numerals you just add the roman numerals for 1-9 to X.

Example

10 = X
11 = X + I or XI
14 = X + IV or XIV
17 = X + VII or XVII

The roman numeral for 20 is XX (20 = 10 + 10 or X + X). To write roman numerals for 21 to 29, just add the roman numerals for 1 through 9 to XX.

Example

20 = XX
21 = XX + I or XXI
24 = XX + IV or XXIV
25 = XX + V or XXV

Exercise A

Write roman numerals for the arabic numerals.

1. 11 ....

2. 18 ....

3. 14 ....

4. 16 ....

5. 12 ....

6. 20 ....

7. 13 ....

8. 19 ....

9. 15 ....

10. 17 ....

11. 20 ....

12. 25 ....

13. 21 ....

14. 26 ....

15. 24 ....

16. 29 ....

17. 22 ....

18. 28 ....

Exercise B

Write roman numerals for the arabic numerals.

1. 14 ....

2. 12 ....

3. 17 ....

4. 26 ....

5. 29 ....

6. 28 ....

7. 27 ....

8. 11 ....

9. 16 ....

10. 20 ....

Writing Roman Numerals 30 to 80

Instruction

The roman numeral for 30 is XXX (30 = 10 + 10 + 10 or X + X + X). To write numerals 31 to 39, just add the roman numerals for 1 through 9 to XXX.

Example

30 = XXX
31 = XXX + I or XXXI
37 = XXX + V + I + I or XXXVII
39 = XXX + IX or XXXIX

The Romans used L for 50. Thus, 40 = XL. The smaller number, X, is in front of the larger number, L. As we learned earlier, the smaller number is subtracted from the larger number when it appears in front of the larger number. When a smaller number follows a larger number, it is added to the larger number. LX = 60.

Example

40 = XL or 40 = 50 − 10
70 = LXX or 70 = 50 + 10 + 10

Exercise A   Write roman numerals for the arabic numerals below.

1. 30 _____   2. 36 _____   3. 31 _____

4. 35 _____   5. 38 _____   6. 32 _____

7. 33 _____   8. 37 _____   9. 34 _____

10. 39 _____   11. 40 _____   12. 70 _____

13. 60 _____   14. 50 _____   15. 80 _____

Exercise B   Write roman numerals for the arabic numerals below.

1. 30 _____   2. 50 _____

3. 70 _____   4. 36 _____

5. 35 _____   6. 61 _____

7. 77 _____   8. 40 _____

9. 66 _____   10. 37 _____

Writing Roman Numerals from 90 to 900 and Greater

Instruction

To write 100, the Romans used C. Therefore, 90 = XC.

Example

90 = XC or 90 = 100 − 10
150 = CL or 150 = 100 + 50
300 = CCC or 300 = 100 + 100 + 100

The Romans used D to represent 500. Thus, 400 = CD.

Example

400 = CD or 400 = 500 − 100
700 = DCC or 700 = 500 + 100 + 100

To write 1,000, the Romans used M. Therefore, 900 = CM.

Example

900 = CM or 900 = 1,000 − 100
1,700 = MDCC or 1,700 = 1,000 + 500 + 100 + 100

Exercise A

Use expanded notation to explain the following roman numerals. The first one has been worked for you.

1. 90 = XC ___90 = 100 − 10___   2. 150 = CL _____

3. 350 = CCCL _____   4. 600 = DC _____

5. 800 = DCCC _____   6. 900 = CM _____

7. 1,900 = MCM _____   8. 2,000 = MM _____

Exercise B

Write roman numerals for the following arabic numerals. The first one has been done. The last three are dates.

1. 1,986 MCMLXXXVI   2. 299 _____   3. 356 _____

4. 501 _____   5. 613 _____   6. 950 _____

7. 1,230 _____   8. 755 _____   9. 2,000 _____

10. 2,001 _____   11. 517 _____   12. 98 _____

13. 1900 _____   14. 1944 _____   15. 1812 _____

UNIT 3—PRETEST
FRACTIONS

Write the fractions described below.

1. numerator 3, denominator 23 _____   2. denominator 17, numerator 2 _____

3. numerator 40, denominator 40 _____   4. denominator 22, numerator 11 _____

Identify each of the following as a proper fraction, an improper fraction, or a mixed number.

5. $\frac{6}{7}$ --   6. $\frac{101}{16}$ --------------------------------------

7. $7\frac{1}{4}$ --   8. $\frac{9}{11}$ --

Determine which fractions can be reduced and reduce them to lowest terms.

9. $\frac{1}{2}$ --   10. $\frac{3}{33}$ ---------------------------------------

11. $\frac{3}{9}$ --   12. $\frac{5}{40}$ ---------------------------------------

Reduce these improper fractions.

13. $\frac{99}{10}$ --   14. $\frac{20}{3}$ --------------------------------------

15. $\frac{76}{6}$ ---   16. $\frac{11}{11}$ -------------------------------------

Read and solve these word problems. Blacken the letter to the right that corresponds to the correct answer.

17. Of the 24 students in Ms. Martinez's class, 4 turned in their projects early. What fraction of the class did not turn them in early?

    a. $\frac{5}{6}$ b. $\frac{2}{3}$ c. $\frac{1}{3}$ d. $\frac{1}{4}$ [a] [b] [c] [d]

18. Bill Herron sold 11 paintings last year. Seven of them were watercolors and the rest were oil paintings. What fraction of the works were watercolor paintings?

    a. $\frac{5}{7}$ b. $\frac{7}{8}$ c. $\frac{7}{11}$ d. $\frac{2}{3}$ [a] [b] [c] [d]

19. Yvonne Davis worked 8 hours less than her normal 40-hour week. What fraction of the week did she work?

    a. $\frac{2}{3}$ b. $\frac{4}{5}$ c. $\frac{5}{6}$ d. $\frac{5}{8}$ [a] [b] [c] [d]

FRACTIONS

Writing Fractions

Instruction   A fraction is an expression which has a bottom number and a top number. The top number is the **numerator**. The bottom number is the **denominator**.

Example

$$\frac{2}{3} \longleftarrow \text{top} \qquad \frac{2}{3} \longleftarrow \text{numerator}$$
$$\frac{2}{3} \longleftarrow \text{bottom} \qquad \frac{2}{3} \longleftarrow \text{denominator}$$

Exercise A   Write the fractions described below.

1. numerator 4, denominator 5 ....

2. numerator 1, denominator 6 ....

3. numerator 3, denominator 8 ....

4. denominator 100, numerator 5 ....

5. denominator 3, numerator 4 ....

6. denominator 98, numerator 18 ....

7. denominator 33, numerator 23 ....

8. denominator 4, numerator 4 ....

SPECIAL *Sale*
BIRTHSTONE AND OPAL
RINGS
HUNDREDS TO CHOOSE FROM
1/2 PRICE
SALE INCLUDES MEN'S AND
LADIES RINGS...BUY SEVERAL NOW!

Exercise B   Write the fractions described below.

1. numerator 28, denominator 73 ....

2. numerator 1, denominator 5 ....

3. denominator 400, numerator 25 ....

4. denominator 3, numerator 96 ....

5. denominator 50, numerator 50 ....

6. denominator 18, numerator 4 ....

7. denominator 23, numerator 21 ....

8. numerator 9, denominator 3 ....

LESSON TWO: Writing Fractions

Instruction

The denominator tells how many parts it takes to make a whole amount. The numerator tells how many parts of the whole the fraction is dealing with.

Example

There are 24 hours in a day.
A carpenter works 8 hours a day.

$\frac{8}{24}$ parts
parts make a whole

There are 10 people in your family.
Today, all the family goes to church.

$\frac{10}{10}$ parts
parts make a whole

There are two pies on the table. Each one is cut into 6 pieces. The family eats 8 pieces at supper.

$\frac{8}{6}$ parts eaten
parts make a whole pie

Exercise A

Read each problem and write the fraction described.

1. Joel plowed 32 rows for a garden. He planted 16 rows of corn. What fraction of the garden has he planted in corn? _____

2. Your child attends kindergarten 5 days a week. She missed one day of class this week. What fraction of the week has she missed? _____

3. On a baseball team, 4 of the starting 9 players have a .325 batting average. What fraction of the starters have a .325 average? _____

Exercise B

Read each problem and write the fraction described.

1. Of the 6 people in Mary's family, she is the only one who likes to play tennis. What part of Mary's family likes to play tennis? _____

2. Earline bought a 10-ounce cold drink and spilled 3 ounces when opening it. What part of the cold drink did she have left? _____

3. You buy a dozen eggs at the grocery store. When you get home, you notice that 2 of the eggs are broken. What part of the 12 eggs are broken? _____

LESSON THREE: Classifying Proper and Improper Fractions

Instruction

A **proper fraction** has a smaller numerator and a larger denominator.

Example

$\frac{4}{5}$ ⟶ $\frac{\text{smaller}}{\text{larger}}$ ⟶ **Proper Fraction**

An **improper fraction** has a larger numerator and a smaller denominator. Fractions with the same numerator and denominator ($\frac{4}{4}$) are also improper fractions.

Example

$\frac{5}{4}$ ⟶ $\frac{\text{larger}}{\text{smaller}}$ ⟶ **Improper Fraction**

A **mixed number** has a whole number part and a fractional part.

Example

$3\frac{2}{5}$ ⟶ whole number + fraction ⟶ **Mixed Number**

Exercise A

Identify each of the following as a proper fraction, an improper fraction, or a mixed number. Circle the correct words.

1. $\frac{12}{11}$ proper fraction improper fraction mixed number
2. $\frac{7}{12}$ proper fraction improper fraction mixed number
3. $4\frac{1}{4}$ proper fraction improper fraction mixed number
4. $\frac{2}{3}$ proper fraction improper fraction mixed number
5. $\frac{115}{127}$ proper fraction improper fraction mixed number
6. $120\frac{5}{6}$ proper fraction improper fraction mixed number
7. $\frac{387}{9}$ proper fraction improper fraction mixed number

N

Nantck		45	6¼ + ¼
NtGsO	.40b	8	13 + ¼
NtPatnt	.10	284	16¼ − ¼
NHamp	.80	47	47¾ +1¾
NMxAr	.791	12	17⅝ + ¼
NPlnRt	.96	18	14½ + ¼
NProc	1.20e	240	15 + ½
NYTime	.52	650	36⅞
NewbE	.25e	10	5 − ⅛
Newcor	.32	20	13⅜
NwpEl	1.50	52	13⅜ − ⅛
Nichols		98	7¾ + ⅜
Noellnd		5	2⅜ + ⅛
Nolex		28	2⅞ + ⅛
NordR	n	29	12¼ − ⅛
NoCdO	g	33	14⅞ + ⅜
NIPS	pf 4.25	z200	33⅛ +1⅜
NuHrz	n	42	3
NuclDt		55	8⅜ + ⅛
Numac		37	10⅝ + ¼

Exercise B

Circle the correct words to classify the following numbers.

1. $\frac{6}{5}$ proper fraction improper fraction mixed number
2. $\frac{3}{4}$ proper fraction improper fraction mixed number
3. $\frac{8}{12}$ proper fraction improper fraction mixed number
4. $\frac{5}{1}$ proper fraction improper fraction mixed number
5. $2\frac{1}{2}$ proper fraction improper fraction mixed number
6. $\frac{120}{4}$ proper fraction improper fraction mixed number
7. $3\frac{15}{16}$ proper fraction improper fraction mixed number

LESSON FOUR: Reducing Proper Fractions

Instruction

A proper fraction may be reduced by dividing the numerator and denominator **by the same number**.

Example

$$\frac{4}{10} = \frac{4 \div 2}{10 \div 2} = \frac{2}{5}$$

$$\frac{3}{9} = \frac{3 \div 3}{9 \div 3} = \frac{1}{3}$$

$$\frac{6}{8} = \frac{6 \div 2}{8 \div 2} = \frac{3}{4}$$

A fraction is reduced to **lowest terms** when there is no longer a number other than 1 which will divide evenly into both the numerator and denominator.

Example

$$\frac{8}{12} = \frac{8 \div 2}{12 \div 2} = \frac{4}{6} = \frac{4 \div 2}{6 \div 2} = \frac{2}{3}$$

The following flow chart can be used to help you determine if a fraction is in lowest terms.

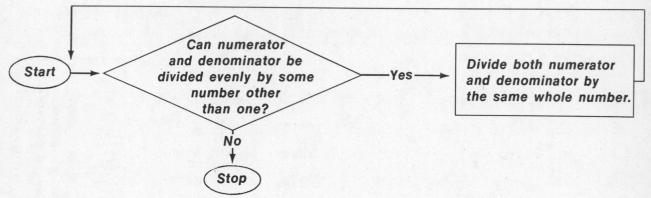

Exercise A

Determine which fractions can be reduced and reduce them to lowest terms.

1. $\frac{5}{10} =$ 2. $\frac{8}{20} =$

3. $\frac{11}{33} =$ 4. $\frac{5}{16} =$

5. $\frac{9}{12} =$ 6. $\frac{12}{24} =$

7. $\frac{22}{44} =$ 8. $\frac{3}{11} =$

9. $\frac{4}{20} =$ 10. $\frac{5}{35} =$

11. $\frac{9}{15} =$ 12. $\frac{6}{66} =$

13. $\frac{18}{72}$ =

14. $\frac{1}{4}$ =

15. $\frac{2}{13}$ =

16. $\frac{13}{39}$ =

17. $\frac{25}{125}$ =

18. $\frac{5}{27}$ =

19. $\frac{6}{24}$ =

20. $\frac{14}{56}$ =

21. $\frac{24}{32}$ =

22. $\frac{9}{30}$ =

23. $\frac{2}{41}$ =

24. $\frac{15}{60}$ =

Exercise B

Determine which fractions can be reduced and reduce them to lowest terms.

1. $\frac{6}{12}$ =

2. $\frac{7}{12}$ =

3. $\frac{3}{15}$ =

4. $\frac{2}{5}$ =

5. $\frac{25}{100}$ =

6. $\frac{13}{36}$ =

7. $\frac{3}{36}$ =

8. $\frac{12}{48}$ =

9. $\frac{15}{65}$ =

10. $\frac{7}{42}$ =

11. $\frac{13}{52}$ =

12. $\frac{16}{28}$ =

13. $\frac{5}{6}$ =

14. $\frac{21}{22}$ =

15. $\frac{24}{100}$ =

16. $\frac{14}{49}$ =

17. $\frac{12}{60}$ =

18. $\frac{7}{27}$ =

19. $\frac{16}{18}$ =

20. $\frac{14}{70}$ =

21. $\frac{2}{3}$ =

22. $\frac{13}{70}$ =

23. $\frac{5}{15}$ =

24. $\frac{15}{75}$ =

25. $\frac{16}{17}$ =

26. $\frac{30}{35}$ =

LESSON FIVE: Reducing Improper Fractions

Instruction

Remember, an improper fraction has a numerator that is greater than or equal to its denominator. An improper fraction is reduced by dividing the numerator by the denominator. Sometimes, the fractional part of a resulting mixed number will need to be reduced to lowest terms.

Example

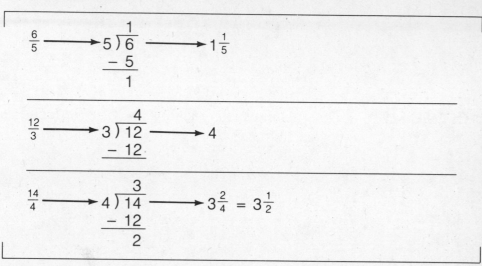

An improper fraction is usually not left as the answer to a problem. An improper fraction should be written as a mixed number in lowest terms.

Exercise A

Reduce each of the following improper fractions. Use long division if necessary.

1. $\frac{7}{2} =$ 2. $\frac{6}{5} =$ 3. $\frac{13}{4} =$

4. $\frac{18}{4} =$ 5. $\frac{21}{7} =$ 6. $\frac{52}{3} =$

7. $\frac{20}{5} =$ 8. $\frac{39}{6} =$ 9. $\frac{100}{3} =$

Exercise B

Reduce these improper fractions.

1. $\frac{5}{3} =$ 2. $\frac{9}{8} =$ 3. $\frac{21}{2} =$

4. $\frac{26}{8} =$ 5. $\frac{45}{9} =$ 6. $\frac{63}{4} =$

7. $\frac{33}{11} =$ 8. $\frac{38}{4} =$ 9. $\frac{200}{3} =$

LESSON SIX:
Writing Improper Fractions for the Whole Number 1 and Reducing Improper Fractions With Same Numerator and Denominator

Instruction

A fraction whose numerator and denominator are the same is equal to 1.

Example

$$\frac{4}{4} = 1 \qquad \frac{9}{9} = 1 \qquad \frac{100}{100} = 1 \qquad \frac{22}{22} = 1$$

Exercise A

Complete the following by filling in each box with the correct number.

1. $\frac{\square}{5} = 1$

2. $\frac{15}{\square} = 1$

3. $\frac{43}{43} = \square$

4. $\frac{1}{1} = \square$

5. $\frac{\square}{8} = 1$

6. $\frac{39}{\square} = 1$

7. $\frac{12}{\square} = 1$

8. $\frac{3}{\square} = 1$

9. $\frac{\square}{51} = 1$

Exercise B

Reduce each of these improper fractions.

1. $\frac{18}{18} =$

2. $\frac{56}{56} =$

3. $\frac{7}{7} =$

4. $\frac{106}{106} =$

5. $\frac{11}{11} =$

6. $\frac{34}{34} =$

7. $\frac{21}{21} =$

8. $\frac{10}{10} =$

9. $\frac{1}{1} =$

10. $\frac{15}{15} =$

11. $\frac{147}{147} =$

12. $\frac{99}{99} =$

Courtesy California Division of Highways

UNIT 3—REVIEW

Instruction

Remember, **proper fractions** have numerators that are smaller than their denominators.

A proper fraction is reduced to lower terms by dividing both the numerator and the denominator by the same number. The fraction is in lowest terms when the numerator and denominator cannot both be evenly divided by the same number other than 1.

An **improper fraction** has a numerator that is equal to or greater than its denominator.

An improper fraction is reduced by dividing the numerator by the denominator. The fractional part of a resulting mixed number will sometimes need to be reduced to lowest terms.

A fraction with the same numerator and denominator is equal to 1.

Directions

Classify each fraction by writing *proper* or *improper* in each blank. Then reduce the fractions. The first one has been worked.

1. _proper_ _____ $\frac{5}{10} = \frac{1}{2}$ 2. _____ $\frac{8}{2} =$

3. _____ $\frac{10}{15} =$ 4. _____ $\frac{4}{12} =$

5. _____ $\frac{15}{4} =$ 6. _____ $\frac{14}{8} =$

7. _____ $\frac{3}{3} =$ 8. _____ $\frac{12}{15} =$

9. _____ $\frac{9}{12} =$ 10. _____ $\frac{20}{4} =$

11. _____ $\frac{33}{33} =$ 12. _____ $\frac{26}{4} =$

13. _____ $\frac{15}{10} =$ 14. _____ $\frac{16}{48} =$

15. _____ $\frac{25}{5} =$ 16. _____ $\frac{22}{7} =$

17. _____ $\frac{75}{100} =$ 18. _____ $\frac{8}{12} =$

19. _____ $\frac{22}{3} =$ 20. _____ $\frac{14}{6} =$

21. _____ $\frac{80}{9} =$ 22. _____ $\frac{36}{3} =$

23. _____ $\frac{9}{21} =$ 24. _____ $\frac{25}{100} =$

Directions
Reduce each expression to lowest terms.

1. $\frac{8}{12} =$

2. $\frac{3}{9} =$

3. $\frac{6}{6} =$

4. $\frac{32}{4} =$

5. $\frac{20}{15} =$

6. $\frac{52}{3} =$

7. $\frac{17}{4} =$

8. $\frac{10}{10} =$

9. $\frac{13}{39} =$

10. $\frac{32}{10} =$

11. $\frac{6}{16} =$

12. $\frac{20}{30} =$

13. $\frac{14}{42} =$

14. $\frac{26}{32} =$

15. $\frac{12}{64} =$

16. $\frac{22}{33} =$

17. $\frac{24}{40} =$

18. $\frac{35}{21} =$

19. $\frac{50}{20} =$

20. $\frac{9}{42} =$

Directions
Solve the word problems. Reduce each answer to lowest terms.

21. The distance from my house to school is 14 miles. I drive 2 miles to the city Park and Ride lot. What part of the distance do I drive? ....

22. Mario lighted 10 candles on Carol's birthday cake. Carol blew out 8 of the candles. What part of the candles did she blow out? ....

23. Mark agreed to baby-sit for 3 hours. The Flemmings were out for 5 hours, and Mark had to work overtime. He worked what part of the contracted time? ....

24. There were 204 people in Elena's class in high school. Twelve of the former classmates became teachers. What part of the class became teachers? ....

25. Marcia regularly runs 2 miles each morning. Today she ran 3 miles. What part of her usual distance did Marcia run this morning? ....

Wide World Photos

29

BUDGETING

Budgeting can be helpful in identifying what part of your income goes for what. A good budget can be very simple. All you must do is make a list of the things for which you spend money. Organize your spending into categories. Keep a record of your spending for a month. You will see a pattern develop. This pattern can be used to make budgets for coming months.

Given below is a monthly budget for a family of four. Perhaps you can adjust these categories to make a budget for your family.

MONTHLY BUDGET

Income
  take-home pay (after deductions) $1,590

Expenses
  mortgage (including principal, interest, insurance, taxes) or rent . . 450
  insurance premiums (life, auto, medical) 90
  household operating expenses (utilities, phone) 190
  transportation (auto loan payment, gasoline) 200
  installment loans (appliances) 44
  food 280
  clothing 100
  recreation 60
  savings 80
  miscellaneous 96

                                                    TOTAL $1,590

Directions

Use the information given in the budget to determine what fractional part of the income is used in the following categories.

	Category	Income	Expenses	Fraction of Income
1.	insurance	$1,590	$90	$\frac{90}{1,590} = \frac{3}{53}$
2.	household expenses			
3.	mortgage			
4.	installment loans			
5.	clothing			

UNIT 4—PRETEST
ADDING FRACTIONS

Add these fractions, reducing to lowest terms. Blacken the letter to the right that corresponds to the correct answer.

1. $\frac{2}{13} + \frac{3}{13} =$ a. $\frac{5}{26}$ b. $\frac{6}{13}$ c. $\frac{5}{13}$ d. $\frac{15}{16}$ [a] [b] [c] [d]

2. $\frac{1}{9} + \frac{1}{9} =$ a. $\frac{2}{18}$ b. $\frac{2}{9}$ c. $\frac{1}{18}$ d. $\frac{9}{9}$ [a] [b] [c] [d]

3. $\frac{2}{5} + \frac{1}{5} =$ a. $\frac{3}{5}$ b. $\frac{2}{10}$ c. $\frac{7}{6}$ d. $\frac{3}{10}$ [a] [b] [c] [d]

4. $\frac{4}{7} + \frac{1}{7} =$ a. $\frac{3}{7}$ b. $\frac{5}{14}$ c. $\frac{5}{7}$ d. $\frac{5}{14}$ [a] [b] [c] [d]

5. $\frac{1}{8} + \frac{1}{8} =$ a. $\frac{2}{8}$ b. $\frac{1}{4}$ c. $\frac{8}{9}$ d. $\frac{2}{16}$ [a] [b] [c] [d]

6. $\frac{2}{9} + \frac{1}{9} =$ a. $\frac{1}{3}$ b. $\frac{3}{18}$ c. $\frac{3}{9}$ d. $\frac{2}{18}$ [a] [b] [c] [d]

7. $\begin{array}{r} \frac{3}{10} \\ \frac{2}{10} \\ + \frac{7}{10} \\ \hline \end{array}$ a. $1\frac{1}{5}$ b. $10\frac{1}{10}$ c. $\frac{12}{10}$ d. 12 [a] [b] [c] [d]

8. $\begin{array}{r} \frac{1}{2} \\ + \frac{1}{8} \\ \hline \end{array}$ a. $\frac{2}{8}$ b. $\frac{1}{4}$ c. $\frac{3}{8}$ d. $\frac{5}{8}$ [a] [b] [c] [d]

9. $\begin{array}{r} \frac{2}{3} \\ + \frac{5}{12} \\ \hline \end{array}$ a. $1\frac{7}{12}$ b. $1\frac{1}{12}$ c. $\frac{7}{12}$ d. $\frac{13}{12}$ [a] [b] [c] [d]

10. $\frac{1}{5} + \frac{3}{15} =$ a. $1\frac{4}{15}$ b. $\frac{4}{20}$ c. $\frac{2}{5}$ d. $\frac{1}{2}$ a   b   c   d

11. $\frac{3}{10} + \frac{2}{5} =$ a. $\frac{5}{15}$ b. $\frac{10}{10}$ c. $1\frac{2}{5}$ d. $\frac{7}{10}$ a   b   c   d

12. $\frac{2}{5} + \frac{2}{3} =$ a. $\frac{4}{5}$ b. $\frac{16}{16}$ c. $1\frac{1}{15}$ d. $1\frac{4}{5}$ a   b   c   d

13. $\frac{1}{3} + \frac{1}{2} =$ a. $\frac{6}{5}$ b. $\frac{5}{6}$ c. $1\frac{1}{6}$ d. $1\frac{1}{5}$ a   b   c   d

14. $+ \frac{3}{4}$ a. $1\frac{1}{28}$ b. $\frac{5}{11}$ c. $\frac{6}{28}$ d. $1\frac{5}{7}$ a   b   c   d

     $+ \frac{2}{7}$

15. $5\frac{1}{4}$ a. $9\frac{1}{4}$ b. $9\frac{1}{2}$ c. $20\frac{1}{2}$ d. 9 a   b   c   d

   $+ 4\frac{1}{4}$

16. $97\frac{1}{3}$ a. $98\frac{1}{3}$ b. 97 c. 98 d. $97\frac{2}{3}$ a   b   c   d

  $+ \frac{1}{3}$

17. $201\frac{1}{16}$ a. $224\frac{1}{4}$ b. $224\frac{1}{8}$ c. $224\frac{1}{2}$ d. 224 a   b   c   d

  $+ 23\frac{3}{16}$

18. $8\frac{3}{8}$ a. $11\frac{5}{8}$ b. $5\frac{1}{4}$ c. $11\frac{4}{8}$ d. $5\frac{1}{8}$ a   b   c   d

  $+ 3\frac{1}{4}$

19. $64\frac{2}{5}$ a. $76\frac{11}{15}$ b. $76\frac{1}{5}$ c. $76\frac{3}{15}$ d. $52\frac{1}{2}$ a   b   c   d

  $+ 12\frac{1}{3}$

20. $5\frac{3}{8}$ a. $15\frac{4}{11}$ b. $14\frac{17}{24}$ c. $15\frac{1}{2}$ d. $14\frac{4}{11}$ a   b   c   d

  $+ 9\frac{1}{3}$

21. $3\frac{2}{5}$ a. $8\frac{9}{10}$ b. $9\frac{1}{10}$ c. $9\frac{11}{10}$ d. $9\frac{9}{10}$ [a] [b] [c] [d]

     $+\ 5\frac{7}{10}$

22. $1\frac{1}{2}$ a. $7\frac{1}{2}$ b. $7\frac{2}{2}$ c. $8\frac{1}{2}$ d. 8 [a] [b] [c] [d]

     $+\ 6\frac{1}{2}$

23. $17\frac{4}{5}$ a. 23 b. 22 c. $22\frac{5}{10}$ d. $23\frac{3}{5}$ [a] [b] [c] [d]

     $+\ 5\frac{1}{5}$

24. $14\frac{5}{12}$ a. $53\frac{8}{12}$ b. $53\frac{1}{6}$ c. $54\frac{1}{6}$ d. $54\frac{1}{12}$ [a] [b] [c] [d]

     $+\ 39\frac{3}{4}$

Solve the word problems. Blacken the letter to the right that corresponds to the correct answer.

25. Mr. Marcos filled the gas tank in his car twice last week, buying $16\frac{7}{10}$ gallons on Monday and $18\frac{3}{10}$ gallons on Friday. How many gallons did he purchase?
a. $34\frac{10}{10}$ b. $35\frac{4}{10}$ c. 35 d. 36 [a] [b] [c] [d]

26. Tony drove $1\frac{3}{4}$ hours before lunch and $\frac{5}{6}$ of an hour after lunch. How many hours did he travel?
a. $2\frac{7}{12}$ b. $3\frac{1}{3}$ c. $2\frac{8}{10}$ d. $2\frac{4}{5}$ [a] [b] [c] [d]

27. The Browns give their cat medication in the following dosages: 1 ounce in the morning, $\frac{2}{3}$ of an ounce at noon, and $\frac{2}{3}$ of an ounce at night. How much medication is the cat given each day?
a. $1\frac{4}{3}$ ounces   b. $1\frac{1}{3}$ ounces   c. $2\frac{2}{3}$ ounces
d. $2\frac{1}{3}$ ounces   [a] [b] [c] [d]

28. The liquid ingredients in a recipe consist of $3\frac{3}{4}$ cups of water and $2\frac{1}{2}$ cups of milk. How much liquid goes into the recipe?
a. $5\frac{3}{6}$ cups   b. $5\frac{1}{2}$ cups   c. $6\frac{1}{8}$ cups   d. $6\frac{1}{4}$ cups   [a] [b] [c] [d]

ADDING FRACTIONS

LESSON ONE: **Adding Fractions With Like Denominators**

Instruction

When two fractions have the same denominator, they are added by adding their numerators. The result is written over the denominator.

Example

$$\frac{1}{5}$$
$$+\ \frac{2}{5}$$
$$\overline{\frac{3}{5}}$$

$$\frac{2}{7}$$
$$+\ \frac{3}{7}$$
$$\overline{\frac{5}{7}}$$

$$\frac{3}{11}$$
$$+\ \frac{5}{11}$$
$$\overline{\frac{8}{11}}$$

Exercise A   Add the following fractions.

1. $\frac{1}{8}$
   $+\ \frac{4}{8}$

2. $\frac{5}{7}$
   $+\ \frac{1}{7}$

3. $\frac{7}{15}$
   $+\ \frac{1}{15}$

4. $\frac{1}{9}$
   $+\ \frac{4}{9}$

5. $\frac{3}{8}$
   $+\ \frac{4}{8}$

6. $\frac{2}{6}$
   $+\ \frac{3}{6}$

7. $\frac{2}{10}$
   $+\ \frac{5}{10}$

8. $\frac{6}{14}$
   $+\ \frac{7}{14}$

Exercise B   Add these fractions.

1. $\frac{7}{12}$
   $+\ \frac{4}{12}$

2. $\frac{2}{5}$
   $+\ \frac{2}{5}$

3. $\frac{4}{27}$
   $+\ \frac{6}{27}$

4. $\frac{5}{18}$
   $+\ \frac{6}{18}$

5. $\frac{4}{13}$
   $+\ \frac{5}{13}$

6. $\frac{13}{31}$
   $+\ \frac{7}{31}$

7. $\frac{7}{14} + \frac{4}{14} =$

8. $\frac{18}{21} + \frac{2}{21} =$

9. $\frac{1}{9} + \frac{1}{9} =$

34

Adding Fractions With Like Denominators and Reducing Answers to Lowest Terms

Instruction

Once fractions are added, the answer should be reduced to lowest terms if it is not already in lowest terms.

Example

$$\begin{array}{r} \frac{5}{12} \\[4pt] \frac{1}{12} \\[4pt] +\ \frac{3}{12} \\[2pt] \hline \frac{9}{12} \end{array} = \frac{9 \div 3}{12 \div 3} = \frac{3}{4} \qquad\qquad \begin{array}{r} \frac{1}{5} \\[4pt] \frac{4}{5} \\[4pt] +\ \frac{2}{5} \\[2pt] \hline \frac{7}{5} \end{array} = 1\frac{2}{5}$$

Remember, proper fractions are reduced by dividing the numerator and denominator by the same number. Improper fractions are reduced by dividing the numerator by the denominator.

Exercise A

Add the following fractions. Write the sums in lowest terms.

1. $\begin{array}{r}\frac{3}{10}\\[4pt]+\ \frac{1}{10}\\[2pt]\hline\end{array}$ 2. $\begin{array}{r}\frac{2}{3}\\[4pt]+\ \frac{1}{3}\\[2pt]\hline\end{array}$ 3. $\begin{array}{r}\frac{5}{16}\\[4pt]+\ \frac{1}{16}\\[2pt]\hline\end{array}$ 4. $\begin{array}{r}\frac{7}{4}\\[4pt]+\ \frac{5}{4}\\[2pt]\hline\end{array}$

5. $\begin{array}{r}\frac{3}{4}\\[4pt]+\ \frac{3}{4}\\[2pt]\hline\end{array}$ 6. $\begin{array}{r}\frac{5}{16}\\[4pt]+\ \frac{7}{16}\\[2pt]\hline\end{array}$ 7. $\begin{array}{r}\frac{7}{20}\\[4pt]+\ \frac{11}{20}\\[2pt]\hline\end{array}$ 8. $\begin{array}{r}\frac{4}{9}\\[4pt]+\ \frac{8}{9}\\[2pt]\hline\end{array}$

Exercise B

Add. Write the sums in lowest terms.

1. $\begin{array}{r}\frac{4}{9}\\[4pt]+\ \frac{7}{9}\\[2pt]\hline\end{array}$ 2. $\begin{array}{r}\frac{1}{5}\\[4pt]\frac{2}{5}\\[4pt]+\ \frac{1}{5}\\[2pt]\hline\end{array}$ 3. $\begin{array}{r}\frac{1}{4}\\[4pt]\frac{3}{4}\\[4pt]+\ \frac{1}{4}\\[2pt]\hline\end{array}$ 4. $\begin{array}{r}\frac{5}{8}\\[4pt]\frac{1}{8}\\[4pt]+\ \frac{2}{8}\\[2pt]\hline\end{array}$

5. $\begin{array}{r}\frac{2}{3}\\[4pt]\frac{2}{3}\\[4pt]+\ \frac{2}{3}\\[2pt]\hline\end{array}$ 6. $\begin{array}{r}\frac{2}{8}\\[4pt]\frac{1}{8}\\[4pt]+\ \frac{4}{8}\\[2pt]\hline\end{array}$ 7. $\begin{array}{r}\frac{5}{12}\\[4pt]\frac{3}{12}\\[4pt]+\ \frac{6}{12}\\[2pt]\hline\end{array}$ 8. $\begin{array}{r}\frac{3}{7}\\[4pt]\frac{1}{7}\\[4pt]+\ \frac{5}{7}\\[2pt]\hline\end{array}$

**Adding Fractions With Unlike
Denominators**

Instruction

Fractions may be added only when they have the same denominator. To add fractions with unlike denominators, you must first find a common denominator.

Example

$\dfrac{3}{4}$

$+ \dfrac{1}{8}$

Do these fractions have the same denominator? No.

$\dfrac{3}{4} =$

$+ \dfrac{1}{8} =$

What is the smallest number into which 4 and 8 will divide evenly?

$\dfrac{3}{4} =$

$+ \dfrac{1}{8} =$

Try the larger denominator. Will 4 divide into 8? Yes. Will 8 divide into 8? Yes. Use 8 as the common denominator.

$\dfrac{3}{4} = \dfrac{6}{8}$

$+ \dfrac{1}{8} = \dfrac{1}{8}$

$\dfrac{7}{8}$

To write $\frac{3}{4}$ as an equivalent fraction with a denominator of 8, follow these steps.
- Divide the new denominator by the old denominator. $8 \div 4 = 2$.
- Then, multiply this answer times the old numerator. $2 \times 3 = 6$.
- Write this answer as the new numerator.

Now find the sums.

Exercise A

Add the following fractions. The lowest common denominator is given.

1. $\dfrac{1}{5} = \dfrac{}{10}$

    $+ \dfrac{1}{10} = \dfrac{}{10}$

2. $\dfrac{1}{2} = \dfrac{}{8}$

    $+ \dfrac{1}{8} = \dfrac{}{8}$

3. $\dfrac{7}{12} = \dfrac{}{12}$

    $+ \dfrac{1}{3} = \dfrac{}{12}$

4. $\dfrac{3}{20} = \dfrac{}{20}$

    $+ \dfrac{4}{5} = \dfrac{}{20}$

5. $\dfrac{2}{5} = \dfrac{}{10}$

    $+ \dfrac{3}{10} = \dfrac{}{10}$

6. $\dfrac{1}{6} = \dfrac{}{12}$

    $+ \dfrac{5}{12} = \dfrac{}{12}$

7. $\dfrac{1}{2} = \dfrac{}{8}$

    $+ \dfrac{3}{8} = \dfrac{}{8}$

8. $\dfrac{2}{3} = \dfrac{}{15}$

    $+ \dfrac{1}{15} = \dfrac{}{15}$

9. $\dfrac{2}{5} = \dfrac{}{10}$

$+\dfrac{3}{10} = \dfrac{}{10}$

10. $\dfrac{5}{6} = \dfrac{}{6}$

$+\dfrac{1}{12} = \dfrac{}{6}$

11. $\dfrac{3}{4} = \dfrac{}{16}$

$+\dfrac{1}{16} = \dfrac{}{16}$

12. $\dfrac{1}{3} = \dfrac{}{9}$

$+\dfrac{1}{9} = \dfrac{}{9}$

13. $\dfrac{3}{15} = \dfrac{}{15}$

$+\dfrac{2}{3} = \dfrac{}{15}$

14. $\dfrac{1}{2} = \dfrac{}{12}$

$+\dfrac{5}{12} = \dfrac{}{12}$

15. $\dfrac{4}{9} = \dfrac{}{9}$

$+\dfrac{1}{3} = \dfrac{}{9}$

16. $\dfrac{3}{16} = \dfrac{}{16}$

$+\dfrac{1}{4} = \dfrac{}{16}$

17. $\dfrac{1}{8} = \dfrac{}{8}$

$+\dfrac{3}{4} = \dfrac{}{8}$

18. $\dfrac{3}{5} = \dfrac{}{10}$

$+\dfrac{1}{10} = \dfrac{}{10}$

19. $\dfrac{3}{7} = \dfrac{}{14}$

$+\dfrac{3}{14} = \dfrac{}{14}$

20. $\dfrac{3}{8} = \dfrac{}{16}$

$+\dfrac{3}{16} = \dfrac{}{16}$

Exercise B   Find the sums.

1. $\dfrac{1}{4} = \dfrac{}{4}$

$+\dfrac{1}{2} = \dfrac{}{4}$

2. $\dfrac{1}{4} = \dfrac{}{16}$

$+\dfrac{1}{16} = \dfrac{}{16}$

3. $\dfrac{3}{5} = \dfrac{}{10}$

$+\dfrac{3}{10} = \dfrac{}{10}$

4. $\dfrac{1}{2} = \dfrac{}{8}$

$+\dfrac{3}{8} = \dfrac{}{8}$

5. $\dfrac{3}{8} = \dfrac{}{16}$

$+\dfrac{5}{16} = \dfrac{}{16}$

6. $\dfrac{1}{8} = \dfrac{}{8}$

$+\dfrac{1}{2} = \dfrac{}{8}$

7. $\dfrac{2}{9} = \dfrac{}{9}$

$+\dfrac{1}{3} = \dfrac{}{9}$

8. $\dfrac{3}{4} = \dfrac{}{8}$

$+\dfrac{1}{8} = \dfrac{}{8}$

9. $\dfrac{2}{9} = \dfrac{}{9}$

$+\dfrac{2}{3} = \dfrac{}{9}$

10. $\dfrac{3}{10} = \dfrac{}{10}$

$+\dfrac{3}{5} = \dfrac{}{10}$

11. $\dfrac{7}{24} = \dfrac{}{24}$

$+\dfrac{5}{12} = \dfrac{}{24}$

12. $\dfrac{2}{21} = \dfrac{}{21}$

$+\dfrac{3}{7} = \dfrac{}{21}$

13. $\dfrac{4}{7} = \dfrac{}{21}$

$+\dfrac{4}{21} = \dfrac{}{21}$

14. $\dfrac{5}{18} = \dfrac{}{18}$

$+\dfrac{1}{3} = \dfrac{}{18}$

15. $\dfrac{1}{15} = \dfrac{}{15}$

$+\dfrac{1}{5} = \dfrac{}{15}$

16. $\dfrac{3}{7} = \dfrac{}{28}$

$+\dfrac{3}{28} = \dfrac{}{28}$

LESSON FOUR: Adding Fractions With Unlike Denominators

Directions Add the following fractions. First find the lowest common denominator.

1. $\dfrac{2}{9}$
   $+\ \dfrac{2}{3}$

2. $\dfrac{1}{5}$
   $+\ \dfrac{4}{15}$

3. $\dfrac{7}{12}$
   $+\ \dfrac{1}{3}$

4. $\dfrac{3}{20}$
   $+\ \dfrac{4}{5}$

5. $\dfrac{2}{7}$
   $+\ \dfrac{4}{21}$

6. $\dfrac{3}{5}$
   $+\ \dfrac{2}{15}$

7. $\dfrac{3}{35}$
   $+\ \dfrac{4}{7}$

8. $\dfrac{2}{3}$
   $+\ \dfrac{1}{6}$

9. $\dfrac{1}{2}$
   $+\ \dfrac{1}{8}$

10. $\dfrac{3}{8}$
   $+\ \dfrac{1}{4}$

11. $\dfrac{1}{2}$
   $+\ \dfrac{3}{8}$

12. $\dfrac{3}{24}$
   $+\ \dfrac{5}{6}$

13. $\dfrac{4}{15}$
   $+\ \dfrac{2}{3}$

14. $\dfrac{3}{7}$
   $+\ \dfrac{3}{14}$

15. $\dfrac{3}{10}$
   $+\ \dfrac{2}{5}$

16. $\dfrac{5}{21}$
   $+\ \dfrac{2}{7}$

17. $\dfrac{1}{10}$
   $+\ \dfrac{1}{5}$

18. $\dfrac{5}{12}$
   $+\ \dfrac{1}{6}$

19. $\dfrac{2}{5}$
   $+\ \dfrac{2}{15}$

20. $\dfrac{5}{24}$
   $+\ \dfrac{1}{12}$

21. $\dfrac{1}{8}$
   $+\ \dfrac{1}{4}$

22. $\dfrac{1}{3}$
   $+\ \dfrac{2}{9}$

23. $\dfrac{3}{8}$
   $+\ \dfrac{1}{2}$

24. $\dfrac{7}{18}$
   $+\ \dfrac{2}{9}$

25. $\dfrac{4}{7}$
   $+\ \dfrac{4}{21}$

26. $\dfrac{3}{4}$
   $+\ \dfrac{1}{16}$

27. $\dfrac{1}{8}$
   $+\ \dfrac{3}{4}$

28. $\dfrac{3}{32}$
   $+\ \dfrac{5}{8}$

Adding Fractions With Unlike Denominators When Neither Is Lowest Common Denominator

Example

$\frac{1}{2}$

$+\ \frac{1}{3}$

The denominators are 2 and 3. 2 will not divide into 3 evenly.

You can always find a common denominator for two fractions by multiplying their denominators together. NOTE: The result will not always be the **lowest** common denominator.

$\frac{1}{2}$

$+\ \frac{1}{3}$

$3 \times 2 = 6$. 6 is a common denominator.

$\frac{1}{2} = \frac{3}{6}$

$+\ \frac{1}{3} = \frac{2}{6}$

$\frac{5}{6}$

Write equivalent fractions with 6 as the denominator. Then find the sum.

Exercise A   Add the following fractions.

1. $\frac{3}{5} = \frac{}{15}$

$+\ \frac{1}{3} = \frac{}{15}$

2. $\frac{1}{2} = \frac{}{14}$

$+\ \frac{1}{7} = \frac{}{14}$

3. $\frac{2}{5} = \frac{}{10}$

$+\ \frac{1}{2} = \frac{}{10}$

4. $\frac{1}{7} = \frac{}{42}$

$+\ \frac{5}{6} = \frac{}{42}$

5. The Jacksons bought two lake lots. One is a $\frac{1}{2}$-acre lot, and the other is $\frac{3}{8}$ of an acre. How much land did they buy?   -----------------

Exercise B   Add these fractions.

1. $\frac{1}{4} = \frac{}{12}$

$+\ \frac{2}{3} = \frac{}{12}$

2. $\frac{2}{7} = \frac{}{28}$

$+\ \frac{1}{4} = \frac{}{28}$

3. $\frac{2}{9} = \frac{}{36}$

$+\ \frac{3}{4} = \frac{}{36}$

4. $\frac{2}{11} = \frac{}{33}$

$+\ \frac{2}{3} = \frac{}{33}$

5. Ana planted strawberries in $\frac{1}{4}$ of her garden space yesterday. Today she planted squash in $\frac{1}{2}$ of the area. How much of her garden has now been planted?   -----------------

39

Adding Fractions With Unlike Denominators When Neither Is Lowest Common Denominator

Directions   Add the following fractions.

1. $\frac{2}{3}$
   $+ \frac{1}{4}$

2. $\frac{3}{4}$
   $+ \frac{1}{5}$

3. $\frac{1}{6}$
   $+ \frac{3}{7}$

4. $\frac{5}{8}$
   $+ \frac{2}{9}$

5. $\frac{3}{10}$
   $+ \frac{2}{3}$

6. $\frac{2}{5}$
   $+ \frac{1}{3}$

7. $\frac{3}{7}$
   $+ \frac{1}{2}$

8. $\frac{2}{5}$
   $+ \frac{1}{6}$

9. $\frac{2}{9}$
   $+ \frac{2}{5}$

10. $\frac{1}{4}$
    $+ \frac{3}{5}$

11. $\frac{3}{7}$
    $+ \frac{1}{4}$

12. $\frac{1}{5}$
    $+ \frac{1}{2}$

13. $\frac{2}{7}$
    $+ \frac{1}{3}$

14. $\frac{1}{4}$
    $+ \frac{1}{7}$

15. $\frac{1}{9}$
    $+ \frac{3}{5}$

16. $\frac{1}{3}$
    $+ \frac{1}{2}$

17. $\frac{1}{4}$
    $+ \frac{1}{5}$

18. $\frac{3}{5}$
    $+ \frac{1}{7}$

19. $\frac{3}{5}$
    $+ \frac{2}{9}$

20. $\frac{2}{3}$
    $+ \frac{3}{11}$

21. $\frac{1}{10}$
    $+ \frac{5}{12}$

22. $\frac{5}{9}$
    $+ \frac{1}{4}$

23. $\frac{1}{3}$
    $+ \frac{1}{7}$

24. $\frac{3}{7}$
    $+ \frac{3}{8}$

25. $\frac{1}{9}$
    $+ \frac{2}{5}$

26. $\frac{2}{11}$
    $+ \frac{5}{9}$

27. $\frac{1}{12}$
    $+ \frac{3}{5}$

28. $\frac{1}{8}$
    $+ \frac{2}{3}$

Adding Mixed Numbers With Like Denominators

Instruction

Remember, a mixed number is a mixture of a whole number and a fraction ($4\frac{3}{4}$, $1\frac{1}{6}$, $120\frac{2}{5}$). To add mixed numbers, add the fractions first. Then add the whole numbers.

Example

$$
\begin{array}{ccc}
2\frac{1}{7} & 2\frac{1}{7} & 2\frac{1}{7} \\
+\ 3\frac{2}{7} & +\ 3\frac{2}{7} & +\ 3\frac{2}{7} \\
& \frac{3}{7} & 5\frac{3}{7}
\end{array}
$$

Exercise A Add the following mixed numbers. Reduce answers to lowest terms.

1. $6\frac{1}{12}$
   $+\ 8\frac{5}{12}$

2. $9\frac{1}{16}$
   $+\ 4\frac{7}{16}$

3. $22\frac{1}{6}$
   $+\ 14\frac{1}{6}$

4. $7\frac{3}{8}$
   $+\ 1\frac{1}{8}$

5. $5\frac{1}{4}$
   $+\ 3\frac{1}{4}$

6. $7\frac{1}{8}$
   $+\ 2\frac{5}{8}$

7. $16\frac{1}{3}$
   $+\ 25\frac{1}{3}$

8. $14\frac{1}{8}$
   $+\ 5\frac{3}{8}$

Exercise B Add. Reduce answers to lowest terms.

1. $8\frac{1}{4}$
   $+\ 3\frac{1}{4}$

2. $3\frac{2}{5}$
   $+\ \frac{1}{5}$

3. $4\frac{3}{11}$
   $+\ 2$

4. $4\frac{1}{10}$
   $+\ \frac{3}{10}$

5. $1\frac{1}{8}$
   $+\ 7\frac{3}{8}$

6. $167\frac{5}{8}$
   $+\ 45$

7. $6\frac{1}{6}$
   $+\ 3\frac{3}{6}$

8. $98\frac{1}{8}$
   $+\ \frac{3}{8}$

9. $\frac{1}{9}$
   $+\ 1\frac{2}{9}$

10. $127\frac{4}{11}$
    $+\ 64\frac{1}{11}$

Adding Mixed Numbers With Unlike Denominators

Instruction

If the fractional parts of the mixed numbers you are adding do not have a common denominator, you must find a common denominator before adding.

Example

$$2\frac{1}{3} \quad\longrightarrow\quad 2\frac{1}{3} = 2\frac{4}{12} \quad\longrightarrow\quad 2\frac{1}{3} = 2\frac{4}{12}$$
$$+\,4\frac{1}{12} \qquad\qquad +\,4\frac{1}{12} = 4\frac{1}{12} \qquad\qquad +\,4\frac{1}{12} = 4\frac{1}{12}$$
$$6\frac{5}{12}$$

$$2\frac{2}{5} \quad\longrightarrow\quad 2\frac{2}{5} = 2\frac{6}{15} \quad\longrightarrow\quad 2\frac{2}{5} = 2\frac{6}{15}$$
$$+\,7\frac{1}{3} \qquad\qquad +\,7\frac{1}{3} = 7\frac{5}{15} \qquad\qquad +\,7\frac{1}{3} = 7\frac{5}{15}$$
$$9\frac{11}{15}$$

Exercise A

Add the following mixed numbers. Reduce answers to lowest terms.

1. $1\frac{3}{8}$
   $+\,1\frac{1}{4}$

2. $3\frac{1}{5}$
   $+\,2\frac{3}{10}$

3. $6\frac{2}{15}$
   $+\,1\frac{3}{5}$

4. $8\frac{1}{4}$
   $+\,4\frac{1}{3}$

5. $3\frac{3}{4}$
   $+\,6\frac{1}{5}$

6. $8\frac{2}{5}$
   $+\,4\frac{1}{3}$

7. $5\frac{2}{3}$
   $+\,6\frac{1}{4}$

8. $3\frac{2}{7}$
   $+\,4\frac{1}{14}$

Exercise B

Add. Reduce answers to lowest terms.

1. $19\frac{2}{11}$
   $+\,4\frac{2}{5}$

2. $8\frac{5}{24}$
   $+\,2\frac{1}{3}$

3. $116\frac{5}{8}$
   $+\,9\frac{1}{24}$

4. $9\frac{4}{5}$
   $+\,2\frac{1}{15}$

5. $24\frac{1}{6}$
   $+\,3\frac{3}{5}$

6. $7\frac{3}{8}$
   $+\,6\frac{1}{3}$

7. $423\frac{1}{12}$
   $+\,16\frac{3}{4}$

8. $6\frac{2}{5}$
   $+\,\frac{7}{20}$

LESSON NINE:

Adding Fractions and Simplifying the Improper Sum

Instruction

If the answer to an addition problem is a whole number mixed with an improper fraction, the answer must be simplified.

$$2\frac{4}{5} = 2\frac{8}{10}$$
$$+ 3\frac{3}{10} = 3\frac{3}{10}$$
$$\overline{5\frac{11}{10}} = 5 + 1\frac{1}{10} = 6\frac{1}{10}$$

$\frac{11}{10}$ is an improper fraction. To simplify the answer to the addition, $\frac{11}{10}$ must be renamed as a mixed number. This mixed number is then added to the whole number part of the answer to the addition problem.

Sometimes, the answer must be reduced so that the fraction will be in lowest terms.

$$5\frac{1}{2} = 5\frac{3}{6}$$
$$+ 7\frac{5}{6} = 7\frac{5}{6}$$
$$\overline{12\frac{8}{6}} = 12 + 1\frac{2}{6} = 13\frac{2}{6} = 13\frac{1}{3}$$

Exercise A

Add. Simplify and reduce answers if necessary.

1. $12\frac{1}{2}$
   $+ 9\frac{7}{10}$

2. $9\frac{5}{8}$
   $+ 6\frac{3}{4}$

3. $2\frac{2}{3}$
   $+ 3\frac{3}{4}$

4. $2\frac{1}{2}$
   $+ 3\frac{1}{2}$

5. $5\frac{3}{7}$
   $+ 2\frac{3}{4}$

6. $2\frac{4}{5}$
   $+ 7\frac{2}{3}$

7. $5\frac{4}{5}$
   $+ 3\frac{2}{3}$

8. $6\frac{4}{9}$
   $+ 3\frac{13}{18}$

Exercise B

Add. Write answers in simplest terms.

1. $3\frac{3}{8}$
   $+ 1\frac{5}{8}$

2. $5\frac{2}{3}$
   $+ 4\frac{1}{2}$

3. $3\frac{4}{5}$
   $+ 6\frac{1}{5}$

4. $15\frac{3}{4}$
   $+ 12\frac{7}{8}$

LESSON TEN: Solving Word Problems

Directions Solve the following word problems. Write answers in simplest terms.

1. A salesperson spends $\frac{1}{10}$ of her salary on gasoline and $\frac{2}{10}$ of her salary on a car payment. What part of her salary does she spend on transportation?

   ----

2. A highway crew completed surfacing $\frac{1}{9}$ of a new highway one week, $\frac{3}{9}$ of the highway the second week, and $\frac{2}{9}$ of the highway the third week. How much of the highway has been completed?

   ----

3. An empty truck weighs $1\frac{1}{2}$ tons. If it is carrying a $\frac{3}{4}$-ton load, what is its total weight?

   ----

4. To make a birthday cake, Lawrence needs $\frac{2}{3}$ cup of sugar for the cake and $2\frac{2}{3}$ cups of sugar for the icing. How much sugar does he need?

   ----

5. Lori bought $4\frac{1}{2}$ pounds of chicken legs, 2 pounds of chicken breasts, and $3\frac{3}{4}$ pounds of chicken livers. How many pounds of chicken did she buy?

   ----

6. Arlene wanted to know how much gasoline was needed to drive to her brother's home in Little Rock. She bought $14\frac{3}{10}$ gallons, $23\frac{7}{10}$ gallons, and $18\frac{9}{10}$ gallons of gasoline. How much gasoline did she buy?

   ----

7. On Friday Rene spent $2\frac{1}{2}$ hours baby-sitting. On Saturday she spent $3\frac{1}{4}$ hours ironing. How many hours did she work in all?

   ----

8. Warren bought $2\frac{1}{12}$ feet of wire, and Adolph bought $4\frac{5}{12}$ feet. How many feet of wire did they buy together?

   ----

9. Maxine bought $3\frac{9}{10}$ pounds of seed for 20¢ a pound and $2\frac{2}{5}$ pounds at 23¢. How many pounds of seed did she buy?

UNIT 4—REVIEW

Directions Add the following fractions. Reduce answers to lowest terms.

1. $\frac{5}{16}$
    $+\ \frac{7}{16}$

2. $\frac{3}{4}$
    $+\ \frac{1}{4}$

3. $\frac{1}{3}$
    $+\ \frac{1}{2}$

4. $\frac{5}{9}$
    $+\ \frac{1}{3}$

5. $9\frac{1}{2}$
    $+\ 15\frac{7}{8}$

6. $8\frac{3}{4}$
    $+\ 12\frac{1}{2}$

7. $17\frac{5}{6}$
    $+\ 12\frac{4}{5}$

8. $6\frac{1}{2}$
    $+\ 7\frac{2}{3}$

9. $9\frac{1}{2}$
    $+\ 8\frac{1}{5}$

10. $7\frac{7}{10}$
    $+\ 14\frac{1}{2}$

11. $\frac{2}{5}$
    $+\ \frac{3}{5}$

12. $\frac{5}{8}$
    $+\ \frac{1}{8}$

13. $\frac{1}{3}$
    $+\ \frac{1}{4}$

14. 7
    $+\ 4\frac{1}{8}$

15. $4\frac{1}{4}$
    $+\ 3\frac{3}{8}$

16. $3\frac{1}{3}$
    $+\ 2\frac{1}{9}$

17. $18\frac{2}{3}$
    $+\ 12\frac{3}{4}$

18. $12\frac{3}{5}$
    $+\ 7\frac{1}{2}$

19. $103\frac{3}{5}$
    $+\ 21\frac{1}{2}$

20. $12\frac{5}{8}$
    $+\ 7\frac{9}{16}$

Directions Study the circle graph which shows the distribution of a store's sales. Answer the questions which follow.

21. What fraction of sales did the hardware department account for? _____

22. How much of the total sales were sporting goods? _____

23. Toys and sporting goods are considered recreational items. What portion of total sales do these items represent? _____

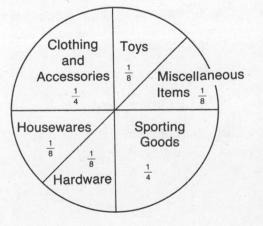

Directions

Solve each problem. Reduce answers to lowest terms.

1. $\dfrac{1}{2}$
$+\ \dfrac{1}{2}$

2. $\dfrac{1}{4}$
$+\ \dfrac{3}{8}$

3. $\dfrac{2}{5}$
$+\ \dfrac{1}{10}$

4. $\dfrac{3}{7}$
$+\ \dfrac{4}{5}$

5. $7\dfrac{1}{12}$
$+\ 3\dfrac{1}{3}$

6. $2\dfrac{1}{2}$
$+\ \dfrac{3}{5}$

7. $15\dfrac{3}{4}$
$+\ 12\dfrac{1}{6}$

8. $13\dfrac{1}{2}$
$+\ 4\dfrac{7}{10}$

9. $4\dfrac{2}{3}$
$+\ 4\dfrac{2}{3}$

10. $6\dfrac{7}{10}$
$+\ 3\dfrac{4}{5}$

11. $12\dfrac{6}{7}$
$+\ 1\dfrac{1}{3}$

12. $3\dfrac{2}{5}$
$+\ 5\dfrac{2}{3}$

13. $4\dfrac{7}{8}$
$+\ \dfrac{1}{2}$

14. $\dfrac{3}{10}$
$+\ 5\dfrac{1}{2}$

15. $6\dfrac{3}{8}$
$+\ 5\dfrac{3}{4}$

16. $12\dfrac{5}{7}$
$+\ 3\dfrac{2}{3}$

17. $8\dfrac{6}{7}$
$+\ 3\dfrac{19}{21}$

18. $4\dfrac{4}{5}$
$+\ 2\dfrac{1}{4}$

19. $5\dfrac{3}{8}$
$+\ 6\dfrac{5}{9}$

20. $13\dfrac{1}{3}$
$+\ 2\dfrac{1}{4}$

Directions

Look at the diagram of Scotsdale Mall and answer these questions.

21. Agnes walks for exercise at Scotsdale Mall. On Monday, she walked around the mall 3 times. How far did she walk? ⎯⎯⎯⎯

22. How many trips around the mall equal one mile? ⎯⎯⎯⎯

23. In one week Agnes walked $\frac{3}{4}$ mile, $1\frac{1}{4}$ miles, 1 mile, $1\frac{1}{2}$ miles, and $1\frac{1}{4}$ miles. How far did she walk in all?

⎯⎯⎯⎯

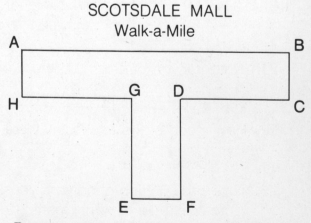

SCOTSDALE MALL
Walk-a-Mile

From any point, a complete trip around the mall equals $\frac{1}{4}$ mile.

LEARNING ABOUT SOCIAL SECURITY

Social Security is a form of government insurance. Most people who have jobs are covered by the program. Your employer deducts a certain amount of your wages from each of your paychecks. This is shown as the F.I.C.A. deduction on your check stub. The employer then matches the deduction and sends the total amount to the government. Self-employed persons must make quarterly payments.

Social security benefits are paid to retired workers and their dependents, to disabled workers and their dependents, and to survivors of workers who have died. Benefits are paid monthly, except for lump-sum death benefits. The amount of social security payments depends upon how long the worker was employed and what the worker's average annual earnings were.

To qualify for benefits, you must have a certain amount of work credits. The exact amount depends upon your age. Each 3-month period that you work and earn $50 or more in pay, equals $\frac{1}{4}$ of a year, or a quarter. If you work a whole calendar year, you receive four quarters of credit. If you are self-employed, you receive four quarters of credit for a year when you have self-employment net profit of $400 or more.

The following tables show how much credit is needed for retirement and survivors benefits. You can get more details at any social security office.

Work credit for retirement benefits

If you reach 62 in	Years you need
1981	7½
1982	7¾
1983	8
1984	8¼
1987	9
1991 or later	10

Work Credit for Survivors Checks

Born after 1929, die at	Born before 1930, die before age 62	Years you need
28 or younger		1½
30		2
32		2½
34		3
36		3½
38		4
40		4½
42		5
44		5½
46		6
48		6½
50		7
52	1981	7½
54	1983	8
56	1985	8½
58	1987	9
60	1989	9½
62 or older	1991 or later	10

Directions

Answer the following questions.

1. Mildred Reeves plans to retire from work in 1989 when she is 62 years old. She has paid into social security while working at different jobs throughout her life. She has accumulated the following work credits.

   1950—$\frac{3}{4}$ yr.   1964—$\frac{1}{2}$ yr.

   1958–61—$2\frac{1}{4}$ yrs.   1968–71—$2\frac{1}{2}$ yrs.

   a. How many years has she contributed to social security? _____

   b. According to the tables, does she have enough credit for retirement?

   _____

2. Lamar Edwards was born in 1933. He died at the age of 49 in 1982, leaving a wife and three children. Mrs. Edwards knows that her husband had social security deductions withheld from his paycheck the following number of years.

   1953—$\frac{1}{4}$ yr.   1960–65—$4\frac{1}{2}$ yrs.

   1955–57—$1\frac{1}{2}$ yrs.   1979—$\frac{3}{4}$ yr.

   a. How many years of work credit did Lamar Edwards have at the time of his

   death? _____

   b. Does his family qualify for social security benefits? _____

3. David Jonas was a self-employed carpenter from 1971 through 1975 and earned more than $400 in each of those years. In all of 1976 and 1977 he worked for a contractor, and FICA was withheld from his paycheck. David plans to retire in 1985 when he reaches 62 years of age. At the time of his retirement, he will have earned $\frac{1}{2}$ year of work credit at his present job at Cannon Industries.

   a. How many years of work credit will David have at the time of his retirement?

   _____

   b. Will he have enough credit for retirement benefits? _____

4. Janice Smith worked for Acme Cleaners from 1955 through 1963. In 1983 she died at the age of 61. In 1957 she did not work for $\frac{1}{2}$ year, and she was also not working for three months in 1961. Acme Cleaners withheld FICA when she was working for them.

   a. How many years of work credit did Janice have when she died? _____

   b. Can her husband collect benefits? _____

UNIT 5—PRETEST
SUBTRACTING FRACTIONS

Subtract the following fractions, reducing answers to lowest terms. Blacken the letter to the right that corresponds to the correct answer.

1. $\frac{3}{5} - \frac{2}{5} =$ a. $\frac{5}{\cdot 5}$ b. $\frac{6}{25}$ c. $\frac{1}{5}$ d. $\frac{1}{0}$ [a] [b] [c] [d]

2. $\frac{8}{9} - \frac{4}{9} =$ a. $\frac{4}{9}$ b. $\frac{5}{9}$ c. $\frac{5}{0}$ d. $1\frac{3}{9}$ [a] [b] [c] [d]

3. $\frac{8}{17}$ a. $\frac{6}{0}$ b. $\frac{10}{17}$ c. $\frac{16}{17}$ d. $\frac{6}{17}$ [a] [b] [c] [d]

   $- \frac{2}{17}$

4. $\frac{6}{7}$ a. $\frac{9}{7}$ b. $\frac{3}{14}$ c. $\frac{3}{7}$ d. 3   [a] [b] [c] [d]

   $- \frac{3}{7}$

5. $\frac{1}{2} - \frac{1}{4} =$ a. $\frac{1}{4}$ b. $\frac{2}{8}$ c. 0   d. $\frac{3}{4}$ [a] [b] [c] [d]

6. $\frac{3}{4} - \frac{1}{16} =$ a. $\frac{2}{16}$ b. $\frac{11}{16}$ c. $\frac{3}{12}$ d. $\frac{3}{16}$ [a] [b] [c] [d]

7. $\frac{2}{3}$ a. $\frac{4}{9}$ b. $\frac{1}{9}$ c. $\frac{0}{9}$ d. $\frac{2}{9}$ [a] [b] [c] [d]

   $- \frac{2}{9}$

8. $\frac{5}{11}$ a. $\frac{3}{22}$ b. $\frac{1}{22}$ c. $\frac{1}{11}$ d. $\frac{3}{11}$ [a] [b] [c] [d]

   $- \frac{4}{22}$

9. $\frac{4}{5} - \frac{2}{3} =$ a. $\frac{6}{8}$ b. $\frac{6}{15}$ c. $\frac{2}{8}$ d. $\frac{2}{15}$ [a] [b] [c] [d]

10. $\frac{10}{21} - \frac{3}{7} =$ a. $\frac{7}{14}$ b. $\frac{1}{21}$ c. $\frac{1}{2}$ d. $\frac{7}{21}$ [a] [b] [c] [d]

11. $\frac{3}{4}$
   $-\frac{1}{3}$

   a. $\frac{4}{7}$ b. $\frac{2}{1}$ c. $\frac{5}{12}$ d. $\frac{7}{12}$ a b c d

12. $\frac{4}{5}$
   $-\frac{1}{2}$

   a. $\frac{3}{10}$ b. $\frac{8}{10}$ c. $\frac{3}{3}$ d. $\frac{3}{7}$ a b c d

13. $5\frac{7}{8} - \frac{1}{4} =$ a. $5\frac{6}{8}$ b. $5\frac{5}{8}$ c. $5\frac{3}{4}$ d. $4\frac{3}{8}$ a b c d

14. $9\frac{11}{12} - \frac{5}{6} =$ a. $9\frac{1}{12}$ b. $9\frac{6}{6}$ c. $10\frac{3}{4}$ d. $9\frac{1}{8}$ a b c d

15. $13\frac{9}{15}$
   $-3\frac{1}{5}$

   a. $16\frac{4}{5}$ b. $10\frac{8}{15}$ c. $11\frac{8}{15}$ d. $10\frac{2}{5}$ a b c d

16. $18\frac{5}{8}$
   $-9\frac{2}{4}$

   a. $9\frac{7}{8}$ b. $9\frac{3}{8}$ c. $9\frac{4}{32}$ d. $9\frac{1}{8}$ a b c d

17. $8 - \frac{3}{10} =$ a. $7\frac{7}{10}$ b. $\frac{5}{10}$ c. $\frac{1}{2}$ d. 7 a b c d

18. $100 - 19\frac{7}{9} =$ a. $80\frac{2}{9}$ b. $71\frac{2}{3}$ c. $71\frac{2}{9}$ d. 80 a b c d

19. 74
   $-10\frac{3}{20}$

   a. $64\frac{17}{20}$ b. $54\frac{17}{20}$ c. $63\frac{17}{20}$ d. $\frac{17}{20}$ a b c d

20. 5
   $-2\frac{2}{6}$

   a. $3\frac{4}{6}$ b. $2\frac{2}{3}$ c. $3\frac{2}{3}$ d. $7\frac{2}{6}$ a b c d

21. $14\frac{1}{3}$
   $-11\frac{3}{4}$

   a. $2\frac{2}{4}$ b. $25\frac{1}{4}$ c. $2\frac{7}{12}$ d. $3\frac{1}{12}$ a b c d

50

22. $8\frac{2}{5}$ a. $1\frac{9}{10}$ b. $1\frac{1}{5}$ c. $\frac{1}{5}$ d. $\frac{9}{10}$ a b c d

    $-\ 7\frac{1}{2}$

23. $97\frac{4}{9}$ a. $12\frac{7}{9}$ b. $11\frac{7}{9}$ c. $12\frac{2}{6}$ d. $11\frac{1}{3}$ a b c d

    $-\ 84\frac{2}{3}$

Solve these word problems. Blacken the letter to the right that corresponds to the correct answer.

24. Kim Lee worked $40\frac{1}{2}$ hours last week and 46 hours this week. How many more hours did he work this week?
a. $6\frac{1}{2}$ b. $4\frac{1}{2}$ c. $5\frac{1}{2}$ d. $4\frac{1}{4}$ a b c d

25. Dora Tijerina usually drives $12\frac{1}{10}$ miles to work each day. Road construction has made it necessary for her to take another route which is actually $\frac{2}{5}$ of a mile shorter. How far does she drive to work each day now?
a. $12\frac{1}{5}$ miles   b. $11\frac{7}{10}$ miles   c. $11\frac{1}{5}$ miles   d. $12\frac{7}{10}$ miles   a b c d

26. Bill made $3\frac{1}{4}$ pounds of fudge and gave $1\frac{1}{2}$ pounds of it to his neighbor. How much candy did he keep for his family?
a. $1\frac{3}{4}$ pounds   b. $1\frac{1}{2}$ pounds   c. $\frac{3}{4}$ pound   d. $2\frac{1}{4}$ pounds   a b c d

27. Roberto bought 10 pounds of grass seed. He used $6\frac{1}{2}$ pounds of the seed on half of his property. How much more seed does he need for the remainder of his property?
a. $3\frac{1}{2}$ pounds   b. $2\frac{1}{2}$ pounds   c. 2 pounds
d. 3 pounds   a b c d

28. Betty and Pam picked $18\frac{3}{4}$ pounds of squash. They want to freeze the squash in packages weighing one pound each. How much squash will be left after the one-pound containers are filled?
a. 17   b. $1\frac{3}{4}$ c. $\frac{3}{4}$ d. $\frac{1}{2}$ a b c d

29. The Adkinsons allow their school-age children to watch television $1\frac{1}{2}$ hours on weekdays and $2\frac{1}{2}$ hours on Saturdays and Sundays. How many hours per week can the children watch television each week?
a. $12\frac{1}{2}$ b. $3\frac{3}{4}$ c. 28   d. $20\frac{1}{2}$ a b c d

SUBTRACTING FRACTIONS

LESSON ONE: **Subtracting Fractions With Like Denominators**

Instruction

When two fractions have the same denominator, you can subtract one numerator from the other numerator and write the result over the denominator.

Example

$$\frac{6}{7}$$
$$-\frac{3}{7}$$
$$\frac{3}{7}$$

$$\frac{5}{13}$$
$$-\frac{2}{13}$$
$$\frac{13}{13}$$

Exercise A Subtract the following fractions.

1. $\frac{5}{9}$
   $-\frac{1}{9}$

2. $\frac{9}{11}$
   $-\frac{2}{11}$

3. $\frac{4}{5}$
   $-\frac{1}{5}$

4. $\frac{2}{3}$
   $-\frac{1}{3}$

5. $\frac{6}{7}$
   $-\frac{2}{7}$

6. $\frac{7}{8}$
   $-\frac{2}{8}$

7. $\frac{9}{13} - \frac{3}{13} =$

8. $\frac{3}{5} - \frac{1}{5} =$

Exercise B Subtract.

1. $\frac{8}{9}$
   $-\frac{3}{9}$

2. $\frac{4}{7}$
   $-\frac{2}{7}$

3. $\frac{4}{16}$
   $-\frac{1}{16}$

4. $\frac{8}{9}$
   $-\frac{4}{9}$

5. $\frac{3}{4}$
   $-\frac{2}{4}$

6. $\frac{10}{17}$
   $-\frac{2}{17}$

7. $\frac{8}{9} - \frac{3}{9} =$

8. $\frac{5}{6} - \frac{4}{6} =$

Oltorf St. EXIT 1/2 MILE

Subtracting Fractions With Unlike Denominators

Instruction

Fractions may be subtracted only when they have the same denominator. To subtract fractions with unlike denominators, you must first find a common denominator. Remember, it is best to use the **lowest** common denominator (the smallest number divisible by both numbers).

Example

$$\frac{3}{4} = \frac{}{4}$$
$$-\frac{1}{2} = \frac{}{4}$$

$$\frac{3}{4} = \frac{3}{4}$$
$$-\frac{1}{2} = \frac{2}{4}$$

$$\frac{3}{4} = \frac{3}{4}$$
$$-\frac{1}{2} = \frac{2}{4}$$
$$\frac{1}{4}$$

Exercise A

Subtract the following fractions. First find the lowest common denominator.

1. $\frac{2}{3}$
   $-\frac{2}{9}$

2. $\frac{14}{15}$
   $-\frac{2}{3}$

3. $\frac{9}{11}$
   $-\frac{9}{22}$

4. $\frac{13}{18}$
   $-\frac{4}{9}$

5. $\frac{13}{25}$
   $-\frac{2}{5}$

6. $\frac{7}{8}$
   $-\frac{1}{4}$

> **LEFTOVER HAM AND BROCCOLI**
>
> 1 pkg (8 oz) medium noodles
> 1½ cups cooked ham, cubed
> 1 can (10¾ oz) cream-of-mushroom soup
> 1 cup milk
> ½ cup sour cream
> 1 pkg (10 oz) broccoli spears, thawed
> 1 cup (4 oz) grated sharp Cheddar cheese
> 1 can (2.8 oz) French-fried onions
> Pimiento
>
> 1. Preheat oven to 350F. Cook noodles according to package directions; drain.

7. $\frac{5}{8} - \frac{1}{2} =$

8. $\frac{1}{3} - \frac{1}{9} =$

Exercise B

Subtract.

1. $\frac{3}{4}$
   $-\frac{1}{2}$

2. $\frac{5}{6}$
   $-\frac{2}{3}$

3. $\frac{3}{8}$
   $-\frac{1}{4}$

4. $\frac{3}{4}$
   $-\frac{5}{16}$

5. $\frac{1}{2}$
   $-\frac{3}{8}$

6. $\frac{3}{4}$
   $-\frac{1}{16}$

7. $\frac{3}{7}$
   $-\frac{1}{14}$

8. $\frac{4}{5}$
   $-\frac{1}{10}$

Subtracting Fractions With Unlike Denominators When Neither Is Lowest Common Denominator

Instruction

Sometimes, you will have to change both denominators to find a common denominator. You can always find a common denominator for two fractions by multiplying their denominators together. The result will not always be the **lowest** common denominator, however.

Example

$$\frac{1}{2} \qquad \frac{1}{2} = \frac{3}{6} \qquad \frac{1}{2} = \frac{3}{6}$$

$$-\frac{1}{3} \qquad -\frac{1}{3} = \frac{2}{6} \qquad -\frac{1}{3} = \frac{2}{6}$$

$$\frac{1}{6}$$

Exercise A

Subtract the following fractions. First find the lowest common denominator.

1. $\frac{3}{4}$

   $-\frac{2}{3}$

2. $\frac{3}{5}$

   $-\frac{1}{3}$

3. $\frac{3}{4}$

   $-\frac{4}{7}$

4. $\frac{2}{3}$

   $-\frac{5}{8}$

5. $\frac{2}{5}$

   $-\frac{3}{8}$

6. $\frac{4}{5}$

   $-\frac{1}{3}$

7. $\frac{7}{9}$

   $-\frac{1}{2}$

8. $\frac{6}{7}$

   $-\frac{1}{3}$

Exercise B

Subtract.

1. $\frac{4}{5}$

   $-\frac{3}{8}$

2. $\frac{7}{9}$

   $-\frac{2}{3}$

3. $\frac{4}{5}$

   $-\frac{8}{11}$

4. $\frac{4}{7}$

   $-\frac{1}{2}$

Read the cake mix directions and answer these questions.

5. The regular directions call for $\frac{2}{3}$ cup of water. How much more water is needed in high altitude areas?

6. The pan is rotated $\frac{1}{4}$ turn three times. If a full turn is $\frac{4}{4}$, how much less than a full turn is the pan moved during baking?

> **HIGH ALTITUDE DIRECTIONS**
> **(3500 to 6500 feet)**
> Preheat oven to 375°. Stir 1 tablespoon all-purpose flour into cake mix (dry). Increase water to ¾ cup.
> **MICROWAVE DIRECTIONS**
> Fold edges of cake pan and prepare cake mix as directed. Cover with waxed paper (paper may stick to top of cake). Microwave on inverted microwavable plate on medium-low (30%) 6 minutes; rotate pan ¼ turn. Microwave on high (100%) 3 minutes, rotating pan ¼ turn every minute. DO NOT TEST FOR DONENESS OR REMOVE WAXED PAPER.

LESSON FOUR:

Subtracting Fractions and Reducing Answers to Lowest Terms

Instruction

Once fractions are subtracted, the answer must be reduced to its lowest term if it is not already in lowest terms. Remember, proper fractions are reduced by dividing the numerator and the denominator **by the same number**.

Example

$$\begin{array}{r} \frac{7}{10} \\ -\ \frac{3}{10} \\ \hline \frac{4}{10} \end{array} = \frac{4 \div 2}{10 \div 2} = \frac{2}{5} \qquad\qquad \begin{array}{r} \frac{3}{4} = \frac{9}{12} \\ -\ \frac{5}{12} = \frac{5}{12} \\ \hline \frac{4}{12} \end{array} = \frac{4 \div 4}{12 \div 4} = \frac{1}{3}$$

Exercise A Subtract the following fractions. Reduce answers to lowest terms.

1. $\begin{array}{r} \frac{1}{3} \\ \frac{1}{4} \\ \hline \end{array}$

2. $\begin{array}{r} \frac{5}{6} \\ -\ \frac{1}{2} \\ \hline \end{array}$

3. $\begin{array}{r} \frac{7}{10} \\ -\ \frac{1}{2} \\ \hline \end{array}$

4. $\begin{array}{r} \frac{5}{6} \\ -\ \frac{3}{4} \\ \hline \end{array}$

5. $\begin{array}{r} \frac{1}{2} \\ -\ \frac{1}{5} \\ \hline \end{array}$

6. $\begin{array}{r} \frac{4}{5} \\ -\ \frac{1}{3} \\ \hline \end{array}$

7. $\begin{array}{r} \frac{6}{9} \\ -\ \frac{1}{3} \\ \hline \end{array}$

8. $\begin{array}{r} \frac{9}{12} \\ -\ \frac{2}{8} \\ \hline \end{array}$

9. $\frac{3}{4} - \frac{1}{8} =$

10. $\frac{7}{9} - \frac{1}{3} =$

11. $\frac{6}{12} - \frac{1}{4} =$

Exercise B Subtract. Write answers in lowest terms.

1. $\begin{array}{r} \frac{7}{12} \\ -\ \frac{1}{3} \\ \hline \end{array}$

2. $\begin{array}{r} \frac{9}{10} \\ -\ \frac{1}{2} \\ \hline \end{array}$

3. $\begin{array}{r} \frac{3}{10} \\ -\ \frac{1}{10} \\ \hline \end{array}$

4. $\begin{array}{r} \frac{3}{4} \\ -\ \frac{1}{2} \\ \hline \end{array}$

5. $\frac{5}{8} - \frac{3}{8} =$

6. $\frac{1}{2} - \frac{3}{8} =$

7. $\frac{5}{9} - \frac{2}{9} =$

8. $\frac{3}{4} - \frac{7}{12} =$

9. $\frac{9}{16} - \frac{3}{8} =$

10. $\frac{3}{4} - \frac{1}{4} =$

Subtracting Mixed Numbers

Instruction

To subtract mixed numbers, subtract the fractions first. Then subtract the whole numbers.

Example

$$2\frac{2}{6} \qquad\qquad 2\frac{2}{6} \qquad\qquad 2\frac{2}{6}$$
$$-1\frac{1}{6} \qquad\qquad -1\frac{1}{6} \qquad\qquad -1\frac{1}{6}$$
$$\qquad\qquad\qquad \frac{1}{6} \qquad\qquad 1\frac{1}{6}$$

$$5\frac{3}{4} \qquad\qquad 5\frac{3}{4} = 5\frac{9}{12} \qquad\qquad 5\frac{3}{4} = 5\frac{9}{12}$$
$$-2\frac{1}{3} \qquad\qquad -2\frac{1}{3} = 2\frac{4}{12} \qquad\qquad -2\frac{1}{3} = 2\frac{4}{12}$$
$$\qquad\qquad\qquad \frac{5}{12} \qquad\qquad 3\frac{5}{12}$$

Exercise A

Subtract the following mixed numbers.

1. $15\frac{3}{4}$ $-\frac{1}{2}$

2. $7\frac{5}{8}$ $-\frac{1}{4}$

3. $26\frac{1}{4}$ $-18\frac{1}{5}$

4. $12\frac{3}{4}$ $-7\frac{1}{2}$

5. $7\frac{7}{10}$ $-4\frac{2}{5}$

6. $823\frac{4}{5}$ $-120\frac{1}{3}$

7. $19\frac{3}{5}$ $-6\frac{1}{3}$

8. $9\frac{2}{3}$ $-5\frac{4}{7}$

Exercise B

Subtract.

1. $7\frac{1}{2}$ $-\frac{1}{4}$

2. $8\frac{3}{8}$ $-4\frac{1}{4}$

3. $12\frac{7}{8}$ $-3\frac{1}{4}$

4. $126\frac{2}{3}$ $-14\frac{1}{2}$

5. $68\frac{3}{8} - 17\frac{1}{4} =$

6. $98\frac{7}{9} - 16\frac{2}{3} =$

7. $223\frac{3}{4} - 167\frac{3}{8} =$

8. $11\frac{4}{5} - \frac{3}{8} =$

Subtracting Mixed Numbers and Reducing Answers to Lowest Terms

Instruction

When you subtract one fraction from another fraction, the answer should be reduced to lowest terms unless it is already in lowest terms.

Remember, a proper fraction is in lowest terms when no number other than 1 will divide evenly into both the numerator and the denominator.

Example

$$16\frac{2}{3} \qquad 16\frac{2}{3} = 16\frac{4}{6} \qquad 16\frac{2}{3} = 16\frac{4}{6}$$
$$-\ 3\frac{1}{6} \qquad -\ 3\frac{1}{6} =\ 3\frac{1}{6} \qquad -\ 3\frac{1}{6} =\ 3\frac{1}{6}$$
$$\qquad\qquad\qquad\qquad\qquad\qquad\qquad 13\frac{3}{6} = 13\frac{1}{2}$$

Exercise A Subtract. Reduce answers to lowest terms.

1. $18\frac{11}{15}$
   $-\ 10\frac{8}{15}$

2. $15\frac{18}{25}$
   $-\ 6\frac{13}{25}$

3. $9\frac{5}{12}$
   $-\frac{1}{6}$

4. $84\frac{7}{12}$
   $-\ 23\frac{1}{4}$

5. $11\frac{2}{3}$
   $-\ 5\frac{7}{15}$

6. $8\frac{4}{50}$
   $-\ 3\frac{3}{100}$

7. $6\frac{9}{12}$
   $-\ 3\frac{1}{4}$

8. $8\frac{5}{6}$
   $-\frac{1}{2}$

Exercise B Subtract. Reduce answers to lowest terms.

1. $21\frac{6}{16}$
   $-\ 3\frac{1}{8}$

2. $5\frac{2}{3}$
   $-\ 1\frac{2}{5}$

3. $19\frac{18}{24}$
   $-\ 13\frac{1}{6}$

4. $33\frac{2}{4}$
   $-\ 21\frac{1}{5}$

5. $1\frac{2}{3}$
   $-\frac{1}{3}$

6. $15\frac{5}{6}$
   $-\ 7\frac{2}{3}$

7. $31\frac{6}{7}$
   $-\ 8\frac{6}{7}$

8. $21\frac{5}{8}$
   $-\ 19\frac{2}{4}$

REGULAR LEADED
1.04 9/10

REGULAR UNLEADED
1.12 9/10

LESSON SEVEN: Subtracting Fractions When Borrowing Is Necessary

Instruction

When subtracting a fraction or a mixed number from a whole number, first change the whole number into a mixed number. Remember, a fraction whose numerator and denominator are the same is equal to 1 ($\frac{5}{5} = 1$, $\frac{16}{16} = 1$, $\frac{1}{1} = 1$).

Example

$$\begin{array}{r} 6 \\ -\,2\frac{2}{7} \\ \hline \end{array} \qquad \begin{array}{rcl} 6 & = & 5\frac{7}{7} \\ -\,2\frac{2}{7} & = & 2\frac{2}{7} \\ \hline & & 3\frac{5}{7} \end{array} \qquad \begin{array}{r} 23 \\ -\,5\frac{5}{19} \\ \hline \end{array} \qquad \begin{array}{rcl} 23 & = & 22\frac{19}{19} \\ -\,5\frac{5}{19} & = & 5\frac{5}{19} \\ \hline & & 17\frac{14}{19} \end{array}$$

Exercise A

Subtract. Be sure answers are in lowest terms.

1. $\begin{array}{r} 14 \\ -\,6\frac{4}{6} \\ \hline \end{array}$

2. $\begin{array}{r} 4 \\ -\,2\frac{2}{5} \\ \hline \end{array}$

3. $\begin{array}{r} 8 \\ -\,4\frac{4}{9} \\ \hline \end{array}$

4. $\begin{array}{r} 21 \\ -\,8\frac{2}{10} \\ \hline \end{array}$

5. $\begin{array}{r} 1 \\ -\,\frac{1}{3} \\ \hline \end{array}$

6. $\begin{array}{r} 15 \\ -\,7\frac{2}{3} \\ \hline \end{array}$

7. $31 - 8\frac{6}{7} =$

8. $21 - 19\frac{2}{4} =$

9. Ione has 5 yards of fabric. Her pattern calls for $2\frac{3}{8}$ yards. How much fabric will she have left? ________________

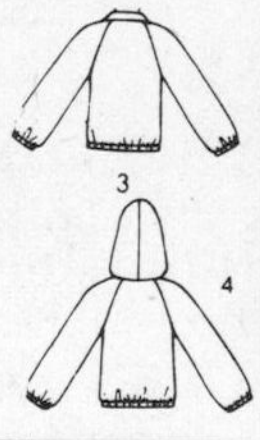

1" wide		2¾	2¾
Elastic for leg (opt.)–⅞ yd. of ¾" wide			
View 2 or 4 Shorts	58" / 60"**	½	⅝
Double fold bias tape–2⅞ yds. of ½" wide			
View 1, 2, 3 or 4 Elastic for pants or shorts waistline–1⅛			
View 1 Top	58" / 60"**	1½	1⅝
View 2 Top	58" / 60"**	1⅜	1⅜
Single fold bias tape–2¼ yds. of 1" wide			
View 3 Jacket	58" / 60"**	1¾	1¾
Interfacing–¼ yd. of 22", 23" or 25" fusible			
Cording–2 yds. of ½" wide			
View 4 Jacket	58" / 60"**	2	2⅛

Exercise B

Subtract. Be sure answers are in lowest terms.

1. $\begin{array}{r} 100 \\ -\,15\frac{1}{9} \\ \hline \end{array}$

2. $\begin{array}{r} 1 \\ -\,\frac{3}{10} \\ \hline \end{array}$

3. $\begin{array}{r} 47 \\ -\,10\frac{4}{12} \\ \hline \end{array}$

4. $\begin{array}{r} 18 \\ -\,4\frac{3}{20} \\ \hline \end{array}$

5. $\begin{array}{r} 69 \\ -\,18\frac{3}{7} \\ \hline \end{array}$

6. $\begin{array}{r} 31 \\ -\,4\frac{3}{6} \\ \hline \end{array}$

7. $601 - 86\frac{9}{10} =$

**Subtracting Fractions When Borrowing Is
Necessary**

Instruction   Sometimes, before subtracting one fraction from another fraction,
borrowing is necessary.

Example

$6\frac{1}{7}$

$-2\frac{2}{7}$

$\frac{2}{7}$ cannot be subtracted from $\frac{1}{7}$. We must borrow from the 6.

$6\frac{1}{7} = 5\frac{8}{7}$

$-2\frac{2}{7} = 2\frac{2}{7}$

Rename the 6 in $6\frac{1}{7}$ as $5\frac{7}{7}$. Then add $5\frac{7}{7} + \frac{1}{7}$.

$6\frac{1}{7} = 5\frac{8}{7}$

$-2\frac{2}{7} = 2\frac{2}{7}$

$3\frac{6}{7}$

Subtract.

$9\frac{2}{5}$

$-3\frac{4}{5}$

$\frac{4}{5}$ cannot be subtracted from $\frac{2}{5}$. We must borrow from the 9.

$9\frac{2}{5} = 8\frac{7}{5}$

$-3\frac{4}{5} = 3\frac{4}{5}$

Rename the 9 in $9\frac{2}{5}$ as $8\frac{5}{5}$. Then add $8\frac{5}{5} + \frac{2}{5}$.

$9\frac{2}{5} = 8\frac{7}{5}$

$-3\frac{4}{5} = 3\frac{4}{5}$

$5\frac{3}{5}$

Subtract.

Exercise A   Subtract these fractions.

1. $8\frac{1}{3}$

$-4\frac{2}{3}$

2. $4\frac{1}{5}$

$-2\frac{3}{5}$

3. $5\frac{2}{7}$

$-1\frac{3}{7}$

4. $12\frac{5}{9}$

$-10\frac{7}{9}$

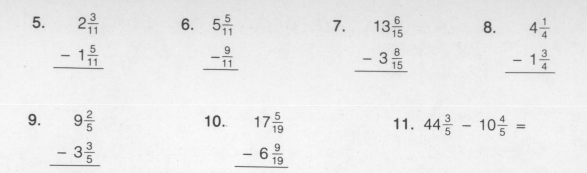

5. $2\frac{3}{11}$
$-1\frac{5}{11}$

6. $5\frac{5}{11}$
$-\frac{9}{11}$

7. $13\frac{6}{15}$
$-3\frac{8}{15}$

8. $4\frac{1}{4}$
$-1\frac{3}{4}$

9. $9\frac{2}{5}$
$-3\frac{3}{5}$

10. $17\frac{5}{19}$
$-6\frac{9}{19}$

11. $44\frac{3}{5} - 10\frac{4}{5} =$

12. Arturo had 2 packages of ground beef weighing $2\frac{1}{4}$ pounds and $1\frac{3}{4}$ pounds. His recipe requires $2\frac{3}{4}$ pounds of meat. If he uses all of the larger package, how much beef will he need from the smaller package?   ------------------

How much ground beef will be left?   ------------------

Exercise B Subtract.

1. $20\frac{7}{15}$
$-10\frac{8}{15}$

2. $3\frac{1}{3}$
$-1\frac{2}{3}$

3. $2\frac{7}{19}$
$-\frac{9}{19}$

4. $56\frac{2}{5}$
$-20\frac{4}{5}$

5. $7\frac{2}{11}$
$-4\frac{3}{11}$

6. $15\frac{2}{7}$
$-14\frac{5}{7}$

7. $9\frac{2}{13}$
$-1\frac{7}{13}$

8. $60\frac{1}{3}$
$-38\frac{2}{3}$

9. $8\frac{2}{5} - 3\frac{4}{5} =$

10. $45\frac{1}{11} - 44\frac{9}{11} =$

11. $38\frac{5}{9} - 3\frac{7}{9} =$

12. $200\frac{2}{7} - 98\frac{3}{7} =$

13. In assembling materials for her hobby, Willa found that she had $16\frac{1}{8}$ yards of blue yarn, $8\frac{3}{8}$ yards of green yarn, and $10\frac{1}{8}$ yards of red yarn. She will use a mixture of colors on 2 projects. One project requires 15 yards of yarn. The other requires $24\frac{1}{4}$ yards. How much more yarn is needed for the 2 projects?   ------------------

60

LESSON NINE: Subtracting Mixed Numbers With Unlike Denominators When Borrowing Is Necessary

Instruction

Remember, before one fraction can be subtracted from another fraction, both fractions must have the same denominator. Be sure the fractions have the same denominator before deciding if it is necessary to borrow.

Example

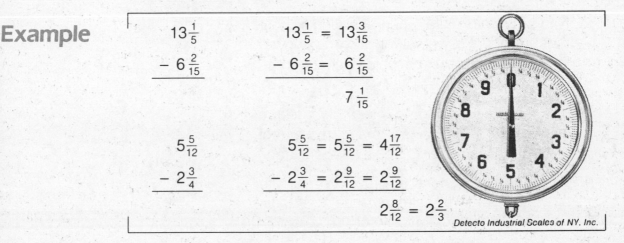

$$13\frac{1}{5} \qquad 13\frac{1}{5} = 13\frac{3}{15}$$
$$-6\frac{2}{15} \qquad -6\frac{2}{15} = 6\frac{2}{15}$$
$$\overline{\qquad} \qquad \overline{\qquad 7\frac{1}{15}}$$

$$5\frac{5}{12} \qquad 5\frac{5}{12} = 5\frac{5}{12} = 4\frac{17}{12}$$
$$-2\frac{3}{4} \qquad -2\frac{3}{4} = 2\frac{9}{12} = 2\frac{9}{12}$$
$$\overline{\qquad} \qquad \overline{\qquad 2\frac{8}{12} = 2\frac{2}{3}}$$

Detecto Industrial Scales of NY. Inc.

Exercise A

Subtract the following fractions. Find the common denominator and borrow if necessary.

1. $9\frac{3}{8}$
   $-7\frac{1}{2}$

2. $7\frac{1}{16}$
   $-4\frac{1}{8}$

3. $15\frac{1}{6}$
   $-12\frac{1}{3}$

4. $25\frac{1}{8}$
   $-12\frac{1}{4}$

5. $20\frac{5}{9}$
   $-16\frac{2}{3}$

6. $19\frac{1}{8}$
   $-14\frac{1}{2}$

Exercise B

Subtract.

1. $6\frac{3}{5}$
   $-4\frac{1}{2}$

2. $9\frac{1}{4}$
   $-7\frac{3}{8}$

3. $15\frac{3}{4}$
   $-12\frac{7}{8}$

4. $37\frac{3}{16}$
   $-18\frac{1}{2}$

5. $8\frac{1}{12}$
   $-3\frac{3}{4}$

6. $12\frac{4}{21}$
   $-5\frac{3}{7}$

LESSON TEN: Solving Word Problems

Directions

Read and solve each problem.

1. Theresa had $\frac{3}{4}$ gallon of milk. She used $\frac{1}{2}$ gallon to make pies for a bazaar. How much milk does she have left?

   ----

2. Naomi is paid extra when she works more than 40 hours a week. For how many overtime hours will she be paid if she worked $56\frac{1}{4}$ hours this week?

   ----

3. The river has risen to $44\frac{1}{2}$ feet. Flood stage level is $52\frac{2}{3}$ feet. How many more feet must the river rise to reach flood stage?

   ----

4. Victor needs $3\frac{1}{15}$ tons of gravel to mix with his cement. If he already has $2\frac{2}{5}$ tons, how much more gravel does he need?

   ----

5. Joan had $\frac{12}{13}$ yard of material. She used $\frac{9}{13}$ yard to make a blouse. How much does she have left?

   ----

6. Chen bought 3 loaves of sourdough bread on sale. He put $2\frac{1}{4}$ loaves in the freezer. How much of a loaf did he keep out?

   ----

7. Easy Finance Company will give you a car loan at $17\frac{1}{2}$% interest. First National Bank's interest rate is $15\frac{3}{4}$%. What is the difference in the interest rates?

   ----

8. One service station advertises regular gasoline for $\$1.12\frac{9}{10}$ per gallon and unleaded gasoline for $\$1.17\frac{9}{10}$. The unleaded is how much more expensive per gallon?

   ----

9. There are $130\frac{2}{5}$ square inches of aluminum foil in Easy Wrap and $125\frac{2}{3}$ square inches in Foil Proof. How many more square inches do you get in Easy Wrap?

   ----

10. There are $12\frac{3}{8}$ ounces in the Choco Chip package and $13\frac{1}{8}$ ounces in the package of Nuts 'N Chips. How many more ounces do you get in a package of Nuts 'N Chips?

UNIT 5—REVIEW

Remember these things when subtracting fractions.

- Fractions can be subtracted only when they have the same denominators.

- When subtracting mixed numbers, subtract the fractions. Then subtract the whole numbers.

- After finding a common denominator, it is sometimes necessary to borrow from the whole number before you can subtract the fractions.

- Be sure to reduce each answer to lowest terms.

Exercise A Subtract the following fractions. Write answers in lowest terms.

1. $\frac{5}{6}$
    $-\frac{1}{6}$

2. $\frac{7}{8}$
    $-\frac{1}{4}$

3. $16\frac{5}{6}$
    $-9\frac{1}{3}$

4. $56\frac{7}{12}$
    $-24\frac{1}{4}$

5. $7\frac{1}{2}$
    $-5\frac{1}{3}$

6. 8
    $-4\frac{1}{2}$

7. $5\frac{1}{3}$
    $-2\frac{2}{3}$

8. $9\frac{1}{3}$
    $-4\frac{5}{6}$

9. $4\frac{3}{4}$
    $-2\frac{5}{6}$

10. $6\frac{1}{3}$
    $-3\frac{1}{2}$

11. $19\frac{6}{7}$
    $-13\frac{1}{4}$

12. $15\frac{1}{3}$
    $-6\frac{4}{6}$

13. $21\frac{5}{12}$
    $-13\frac{1}{4}$

14. $54\frac{6}{21}$
    $-25\frac{3}{7}$

15. 17
    $-1\frac{7}{10}$

Exercise B

Subtract these fractions. Be sure answers are in lowest terms.

1. $\dfrac{7}{18}$
    $-\dfrac{2}{18}$

2. $\dfrac{2}{3}$
    $-\dfrac{1}{3}$

3. $4\dfrac{6}{7}$
    $-1\dfrac{1}{7}$

4. $\dfrac{2}{3}$
    $-\dfrac{1}{4}$

5. $\dfrac{3}{4}$
    $-\dfrac{1}{2}$

6. $\dfrac{5}{6}$
    $-\dfrac{4}{18}$

7. $10\dfrac{4}{5}$
    $-6\dfrac{3}{10}$

8. $4\dfrac{1}{2}$
    $-1\dfrac{2}{5}$

9. $18\dfrac{4}{7}$
    $-12\dfrac{1}{3}$

10. 6
    $-4\dfrac{1}{2}$

11. $3\dfrac{2}{3}$
    -2

12. 5
    $-4\dfrac{3}{4}$

13. $16\dfrac{1}{2}$
    $-5\dfrac{3}{4}$

14. $12\dfrac{3}{5}$
    $-8\dfrac{3}{10}$

15. $156\dfrac{1}{6}$
    $-53\dfrac{1}{2}$

16. $5\dfrac{2}{5}$
    $-1\dfrac{2}{3}$

17. $13\dfrac{1}{8}$
    $-6\dfrac{5}{8}$

18. $24\dfrac{1}{3}$
    $-21\dfrac{3}{4}$

19. $21\dfrac{5}{24}$
    $-9\dfrac{7}{8}$

20. $132\dfrac{9}{10}$
    $-116\dfrac{9}{10}$

SCALE IN MILES

0   1   2   3   4 $\frac{1}{2}$ $\frac{3}{4}$ 1
tenths

21. $18\dfrac{1}{3}$
    -6

22. 5
    $-2\dfrac{6}{7}$

64

READING A STOCK MARKET TABLE

More than 20 million persons own shares in American corporations. A share of stock in a company is a share in the ownership of the company. People invest in stocks to put their money to work for them to earn more money.

The financial sections of many newspapers contain tables showing stock transactions. Look in a newspaper to find a table for the New York Stock Exchange or the American Stock Exchange. The columns labeled high, low, close, and change indicate the following:

high—the highest price of the stock for the day

low—the lowest price of the stock for the day

close—the final price of the stock for the day

change (chg.)—the difference between the closing price for that day and the previous day. The change may be **up** from the previous day (+) or **down** from the previous day (−).

Example

	HIGH	LOW	CLOSE	CHG.
U. S. Steel	$50\frac{1}{4}$	$49\frac{3}{4}$	$49\frac{7}{8}$	$-\frac{1}{8}$

United States Steel sold at a high of $50.25, at a low of $49.75, and closed at $49.875. This reflects a **decrease** of $\frac{1}{8}$ dollar from the previous day's closing. Therefore, to find the closing of the previous day, add $\frac{1}{8}$ to $49\frac{7}{8}$.

$$49\frac{7}{8}$$
$$+ \frac{1}{8}$$
$$\overline{49\frac{8}{8} = 50}$$

If the change reflects an increase (+), subtract the amount of the increase from the closing to find the previous day's closing.

Directions   Look at this reproduction of a small portion of a stock market table. Then answer the questions.

	HIGH	LOW	CLOSE	CHG.
Texaco	$94\frac{3}{4}$	$94\frac{3}{4}$	$94\frac{3}{4}$	$+\frac{1}{8}$
TWA	$96\frac{3}{4}$	$96\frac{1}{2}$	$96\frac{5}{8}$	$+\frac{1}{4}$
UAirL	$55\frac{1}{8}$	$54\frac{3}{4}$	$54\frac{3}{4}$	$-\frac{1}{4}$
U. S. Steel	$68\frac{7}{8}$	$68\frac{1}{2}$	$68\frac{7}{8}$	$+\frac{3}{4}$
Westingh	98	$97\frac{1}{2}$	98	---
Xerox	$102\frac{1}{8}$	$101\frac{1}{2}$	102	$+\frac{1}{2}$

1. What is the difference between the high and low of TWA? ....

2. What was the closing price of TWA on the previous day? ....

3. What was the closing price of United Air Lines (UAirL) on the previous day?

   ....

4. What was the closing price of U. S. Steel on the previous day? ....

5. What was the closing price of Xerox on the previous day? ....

6. What is the difference between the high and low of Texaco? ....

7. What was the closing price of Westinghouse (Westingh) on the previous day?

   ....

8. What is the difference between the high and low of United Air Lines? ....

9. What was the closing price of Texaco on the previous day? ....

10. What is the difference between the high and low of U. S. Steel? ....

11. What is the difference between the high and low of Westinghouse? ....

12. What is the difference between the high and low of Xerox? ....

13. How much higher is Xerox's closing price than TWA's? ....

14. How much lower did United Air Lines close than U. S. Steel? ....

UNIT 6—PRETEST
MULTIPLYING FRACTIONS

Multiply these fractions, reducing answers to lowest terms. Blacken the letter to the right that corresponds to the correct answer.

1. $\frac{1}{3} \times \frac{2}{5} =$ a. $\frac{2}{15}$ b. $\frac{5}{16}$ c. $\frac{5}{6}$ d. $\frac{2}{10}$ ⬚a ⬚b ⬚c ⬚d

2. $\frac{7}{9} \times \frac{1}{2} =$ a. $\frac{14}{9}$ b. $\frac{9}{14}$ c. $\frac{7}{18}$ d. $\frac{6}{7}$ ⬚a ⬚b ⬚c ⬚d

3. $\frac{11}{15} \times \frac{3}{33} =$ a. $\frac{1}{15}$ b. $\frac{1}{5}$ c. $\frac{14}{48}$ d. $\frac{7}{24}$ ⬚a ⬚b ⬚c ⬚d

4. $\frac{1}{2} \times \frac{3}{5} =$ a. $\frac{5}{6}$ b. $\frac{4}{7}$ c. $\frac{3}{3}$ d. $\frac{3}{10}$ ⬚a ⬚b ⬚c ⬚d

5. $17 \times \frac{2}{3} =$ a. $\frac{1}{7}$ b. $11\frac{1}{3}$ c. $\frac{2}{14}$ d. $16\frac{1}{3}$ ⬚a ⬚b ⬚c ⬚d

6. $14 \times \frac{3}{7} =$ a. $14\frac{3}{7}$ b. 6 c. 42 d. $\frac{2}{3}$ ⬚a ⬚b ⬚c ⬚d

7. $\frac{7}{11} \times 22 =$ a. $\frac{11}{22}$ b. $\frac{7}{22}$ c. $\frac{22}{77}$ d. 14 ⬚a ⬚b ⬚c ⬚d

8. $\frac{5}{9} \times 54 =$ a. $10\frac{4}{9}$ b. $28\frac{1}{9}$ c. 30 d. $25\frac{4}{9}$ ⬚a ⬚b ⬚c ⬚d

9. $3\frac{1}{3} \times 3 =$ a. 10 b. 91 c. $6\frac{1}{3}$ d. 4 ⬚a ⬚b ⬚c ⬚d

10. $7\frac{1}{4} \times 12 =$ a. $19\frac{1}{4}$ b. 22 c. $84\frac{1}{4}$ d. 87 ⬚a ⬚b ⬚c ⬚d

11. $1\frac{1}{5} \times 4\frac{2}{3} =$ a. $4\frac{2}{15}$ b. $5\frac{3}{5}$ c. $5\frac{2}{8}$ d. $24\frac{2}{15}$ ⬚a ⬚b ⬚c ⬚d

12. $4 \times 2\frac{6}{7} =$ a. $11\frac{3}{7}$ b. $8\frac{6}{7}$ c. $11\frac{6}{28}$ d. $8\frac{6}{7}$ ⬚a ⬚b ⬚c ⬚d

Solve these word problems. Blacken the letter to the right that corresponds to the correct answer.

13. Joshua and Cara walk $2\frac{1}{2}$ miles after dinner each night. How far do they walk in a week?
    a. $17\frac{1}{2}$ miles   b. $12\frac{1}{2}$ miles   c. $12\frac{1}{4}$ miles   d. 10 miles

    a ☐ b ☐ c ☐ d ☐

14. Roddy bought six $1\frac{2}{5}$-ounce candy bars. He later found that he could have saved 25 cents by buying a package containing 6 of the bars. What is the combined weight of the six bars in the package?
    a. $6\frac{2}{5}$ ounces   b. $6\frac{1}{15}$ ounces   c. $8\frac{2}{5}$ ounces
    d. $7\frac{2}{5}$ ounces

    a ☐ b ☐ c ☐ d ☐

15. Connie recorded some of her favorite television programs on videotape. The shows filled four $4\frac{1}{2}$ hour tapes and two $2\frac{1}{2}$ hour tapes. How many hours of television programs did Connie record?
    a. $23\frac{1}{2}$ b. 22   c. 42   d. 23

    a ☐ b ☐ c ☐ d ☐

16. A roast weighed $7\frac{1}{4}$ pounds when it was bought. After the bone and fat were trimmed away, $6\frac{1}{8}$ pounds of meat remained. How much of the roast was discarded?
    a. $1\frac{1}{4}$ pounds   b. $1\frac{1}{8}$ pounds   c. $\frac{3}{8}$ pound   d. $\frac{1}{4}$ pound

    a ☐ b ☐ c ☐ d ☐

17. Jay's class receives reading instruction from 8:00 to 9:30 Monday through Friday. On Tuesday and Thursday afternoons Jay works on reading skills with a tutor from 4:15 until 5:00. How many hours per week does Jay spend developing his reading skills?
    a. $9\frac{1}{2}$ b. 7   c. $7\frac{1}{2}$ d. 9

    a ☐ b ☐ c ☐ d ☐

MULTIPLYING FRACTIONS

Multiplying Fractions

Instruction To multiply fractions, multiply the numerator times the numerator and the denominator times the denominator.

Example

$$\frac{1}{2} \times \frac{1}{2} \longrightarrow \frac{1}{2} \times \frac{1}{2} = \frac{1}{-} \longrightarrow \frac{1}{2} \times \frac{1}{2} = \frac{1}{4}.$$

$$\frac{2}{3} \times \frac{1}{5} \longrightarrow \frac{2}{3} \times \frac{1}{5} = \frac{2}{-} \longrightarrow \frac{2}{3} \times \frac{1}{5} = \frac{2}{15}$$

Exercise A Multiply these fractions. Write answers in lowest terms.

1. $\frac{1}{3} \times \frac{1}{2} =$ 2. $\frac{1}{2} \times \frac{3}{5} =$ 3. $\frac{2}{3} \times \frac{4}{5} =$

4. $\frac{1}{5} \times \frac{2}{3} =$ 5. $\frac{3}{5} \times \frac{1}{4} =$ 6. $\frac{2}{3} \times \frac{2}{3} =$

7. $\frac{2}{7} \times \frac{3}{5} =$ 8. $\frac{5}{7} \times \frac{3}{4} =$ 9. $\frac{5}{6} \times \frac{1}{4} =$

10. $\frac{2}{3} \times \frac{2}{5} =$ 11. $\frac{1}{6} \times \frac{5}{7} =$

12. $\frac{3}{4} \times \frac{3}{4} =$ 13. $\frac{5}{8} \times \frac{3}{4} =$

14. $\frac{1}{6} \times \frac{3}{4} =$ 15. $\frac{1}{2} \times \frac{1}{2} =$

16. $\frac{3}{10} \times \frac{2}{3} =$ 17. $\frac{2}{3} \times \frac{1}{6} =$

Exercise B Multiply. Reduce answers to lowest terms.

1. $\frac{3}{4} \times \frac{2}{3} =$ 2. $\frac{9}{10} \times \frac{1}{9} =$ 3. $\frac{1}{8} \times \frac{2}{3} =$

4. $\frac{1}{4} \times \frac{1}{8} =$ 5. $\frac{7}{9} \times \frac{3}{7} =$ 6. $\frac{8}{10} \times \frac{2}{3} =$

7. $\frac{1}{9} \times \frac{4}{7} =$ 8. $\frac{2}{5} \times \frac{3}{4} =$ 9. $\frac{5}{6} \times \frac{3}{5} =$

10. $\frac{4}{5} \times \frac{1}{4} =$ 11. $\frac{3}{5} \times \frac{1}{2} =$ 12. $\frac{2}{5} \times \frac{1}{10} =$

LESSON TWO: Using Cancellation To Multiply Fractions

Instruction

Before multiplying fractions, it is sometimes possible to **cancel** and make the problem easier to work.

When you cancel, divide the numerator of one fraction and the denominator of the other fraction **by the same number**. After cancelling, multiply the numerators and multiply the denominators.

Example

$$\frac{3}{4} \times \frac{1}{6} \longrightarrow \frac{\overset{1}{\cancel{3}}}{4} \times \frac{1}{\underset{2}{\cancel{6}}} = \frac{1}{8}$$

The 3 and the 6 were cancelled by dividing each by 3.

Sometimes, you cannot cancel any of the numbers.

Example

$$\frac{2}{3} \times \frac{1}{5} = \frac{2}{15}$$

Sometimes, you can cancel two numbers.

Example

$$\frac{2}{\underset{1}{\cancel{3}}} \times \frac{\overset{2}{\cancel{6}}}{7} = \frac{4}{7}$$

Sometimes, you can cancel all the numbers.

Example

$$\frac{\overset{1}{\cancel{3}}}{\underset{1}{\cancel{4}}} \times \frac{\overset{2}{\cancel{8}}}{\underset{3}{\cancel{9}}} = \frac{2}{3}$$

The 3 and the 9 were cancelled by dividing by 3. The 4 and the 8 were cancelled by dividing by 4.

Exercise A Multiply these fractions. Cancel when possible.

1. $\frac{1}{2} \times \frac{2}{5} =$ 2. $\frac{3}{5} \times \frac{2}{3} =$ 3. $\frac{2}{7} \times \frac{3}{4} =$

4. $\frac{4}{9} \times \frac{3}{5} =$ 5. $\frac{2}{3} \times \frac{3}{4} =$ 6. $\frac{3}{10} \times \frac{5}{6} =$

7. $\frac{1}{3} \times \frac{3}{4} =$ 8. $\frac{5}{9} \times \frac{3}{5} =$ 9. $\frac{9}{10} \times \frac{2}{9} =$

Exercise B Multiply. Cancel when possible.

1. $\frac{11}{15} \times \frac{3}{22} =$ 2. $\frac{1}{8} \times \frac{2}{4} =$ 3. $\frac{1}{2} \times \frac{2}{7} =$

4. $\frac{3}{8} \times \frac{2}{9} =$ 5. $\frac{25}{33} \times \frac{3}{5} =$ 6. $\frac{3}{10} \times \frac{5}{9} =$

Setting Up and Working Problems With Multiplication of Fractions

Instruction

When you see the word *of* in problems dealing with fractions, you need to multiply.

Example

$\frac{1}{2}$ of $\frac{1}{4}$ $\longrightarrow$ $\frac{1}{2} \times \frac{1}{4} = \frac{1}{8}$ Therefore, $\frac{1}{2}$ of $\frac{1}{4}$ is $\frac{1}{8}$.

It is easier to multiply fractions when they are in horizontal form ($\frac{1}{2} \times \frac{1}{2}$) than when they are written in vertical form.

$$\begin{array}{r} \frac{1}{2} \\ \times \frac{1}{2} \\ \hline \end{array}$$

Example

$$\begin{array}{r} \frac{1}{2} \\ \times \frac{1}{2} \\ \hline \end{array} \longrightarrow \frac{1}{2} \times \frac{1}{2} = \frac{1}{4}$$

Exercise A

Write each of the following problems in workable form and multiply. Reduce each answer to lowest terms.

1. $\frac{1}{2}$ of $\frac{3}{4}$ =

2. $\frac{4}{5}$ of $\frac{3}{7}$ =

3.
$$\begin{array}{r} \frac{1}{2} \\ \times \frac{1}{4} \\ \hline \end{array}$$

4.
$$\begin{array}{r} \frac{2}{3} \\ \times \frac{9}{10} \\ \hline \end{array}$$

5.
$$\begin{array}{r} \frac{7}{10} \\ \times \frac{12}{21} \\ \hline \end{array}$$

6.
$$\begin{array}{r} \frac{3}{8} \\ \times \frac{5}{6} \\ \hline \end{array}$$

Exercise B

Write each problem in workable form and multiply. Reduce answers to lowest terms.

1. $\frac{1}{3}$ of $\frac{4}{5}$ =

2. $\frac{2}{3}$ of $\frac{4}{7}$ =

3.
$$\begin{array}{r} \frac{1}{3} \\ \times \frac{2}{5} \\ \hline \end{array}$$

4.
$$\begin{array}{r} \frac{3}{4} \\ \times \frac{4}{9} \\ \hline \end{array}$$

Multiplying a Fraction and a Whole Number

Instruction

When multiplying a fraction and a whole number, first write the whole number as an improper fraction. Then multiply as usual.

Example

$$\frac{1}{2} \times 5 \longrightarrow \frac{1}{2} \times \frac{5}{1} \longrightarrow \frac{1}{2} \times \frac{5}{1} = \frac{5}{2} = 2\frac{1}{2}$$

$$4 \times \frac{3}{5} \longrightarrow \frac{4}{1} \times \frac{3}{5} \longrightarrow \frac{4}{1} \times \frac{3}{5} = \frac{12}{5} = 2\frac{2}{5}$$

Exercise A

Multiply. Reduce answers to lowest terms.

1. $\frac{2}{3} \times 4 =$

2. $\frac{1}{2} \times 9 =$

3. $5 \times \frac{3}{4} =$

4. $4 \times \frac{1}{3} =$

5. $\frac{4}{5} \times 5 =$

6. $\frac{2}{3} \times 3 =$

7. $\frac{7}{8} \times 8 =$

8. $\frac{1}{3} \times 9 =$

9. $100 \times \frac{9}{10} =$

10. $4 \times \frac{5}{7} =$

11. $\frac{1}{9} \times 16 =$

12. $\frac{5}{9} \times 9 =$

13. $4 \times \frac{1}{2} =$

14. $21 \times \frac{1}{8} =$

15. $\frac{3}{8} \times 10 =$

16. $\frac{1}{3} \times 7 =$

Exercise B

Multiply. Reduce answers to lowest terms.

1. $\frac{1}{3} \times 3 =$

2. $\frac{1}{5} \times 5 =$

3. $\frac{9}{10} \times 2 =$

4. $\frac{8}{9} \times 3 =$

5. $\frac{6}{7} \times 14 =$

6. $15 \times \frac{3}{5} =$

7. $\frac{2}{5} \times 11 =$

8. $\frac{1}{8} \times 1 =$

9. $13 \times \frac{2}{3} =$

10. $\frac{9}{11} \times 22 =$

Writing Mixed Numbers as Improper Fractions

Instruction

Follow these steps to change a mixed number into an impropor fraction.

- Multiply the denominator times the whole number.
- Add the numerator to your answer.
- Place the sum over the denominator.

Remember, a proper fraction looks like this: $\frac{4}{5}$. An improper fraction looks like this: $\frac{5}{4}$. A mixed number looks like this: $1\frac{1}{4}$.

Example

Write $3\frac{2}{5}$ as an improper fraction.

1. Multiply the denominator of the fraction times the whole number, 3. $5 \times 3 = 15$.

2. Add the product, 15, to the old numerator of the fraction. $15 + 2 = 17$. This will be the new numerator.

3. Write 17 over the denominator.

$$3\frac{2}{5} = \frac{17}{5}$$

Write $4\frac{5}{6}$ as an improper fraction.

1. Multiply the denominator of the fraction times the whole number, 4. $6 \times 4 = 24$.

2. Add the product, 24, to the old numerator of the fraction. $24 + 5 = 29$. This will be the new numerator.

3. Write 29 over the denominator.

Exercise A

Write each mixed number as an improper fraction.

1. $1\frac{2}{3} =$ 2. $3\frac{3}{4} =$ 3. $5\frac{1}{5} =$

4. $12\frac{2}{3} =$ 5. $45\frac{1}{2} =$ 6. $1\frac{3}{5} =$

Exercise B

Write each mixed number as an improper fraction.

1. $1\frac{4}{5} =$ 2. $4\frac{2}{3} =$ 3. $6\frac{1}{6} =$

4. $13\frac{3}{4} =$ 5. $33\frac{2}{3} =$ 6. $9\frac{1}{4} =$

LESSON SIX:

Writing Improper Fractions as Mixed Numbers

Instruction

Follow these steps to change an improper fraction into a mixed number.

- Divide the numerator by the denominator.
- Write the remainder in fractional form by putting the remainder over the divisor (denominator of the fraction).

Example

Write $\frac{14}{4}$ as a mixed number.

$\frac{14}{4} = 3\frac{2}{4} = 3\frac{1}{2}$ Divide 14 by 4. $14 \div 4 = 3$ with a remainder of 2. Write the remainder in fractional form, $\frac{2}{4}$. Reduce to lowest terms.

Exercise A

Write each improper fraction as a mixed number. Reduce answers to lowest terms.

1. $\frac{10}{3} =$ 2. $\frac{8}{6} =$ 3. $\frac{44}{3} =$

4. $\frac{16}{3} =$ 5. $\frac{9}{8} =$ 6. $\frac{28}{3} =$

7. $\frac{11}{7} =$ 8. $\frac{16}{15} =$ 9. $\frac{27}{3} =$

10. $\frac{5}{2} =$ 11. $\frac{32}{7} =$ 12. $\frac{49}{6} =$

13. $\frac{50}{12} =$ 14. $\frac{14}{12} =$ 15. $\frac{51}{10} =$

Exercise B

Write each improper fraction as a mixed number in lowest terms.

1. $\frac{27}{5} =$ 2. $\frac{12}{8} =$ 3. $\frac{51}{4} =$

4. $\frac{17}{4} =$ 5. $\frac{26}{16} =$ 6. $\frac{46}{9} =$

7. $\frac{19}{4} =$ 8. $\frac{40}{12} =$ 9. $\frac{16}{3} =$

10. $\frac{27}{6} =$ 11. $\frac{57}{14} =$ 12. $\frac{79}{25} =$

13. $\frac{60}{18} =$ 14. $\frac{41}{7} =$ 15. $\frac{35}{3} =$

Multiplying Mixed Numbers

When multiplying with a mixed number, first change the mixed number to an improper fraction. Then multiply numerator times numerator and denominator times denominator.

Example

$$2\frac{3}{4} \times \frac{1}{3} \longrightarrow \frac{11}{4} \times \frac{1}{3} \longrightarrow \frac{11}{4} \times \frac{1}{3} = \frac{11}{12}$$

$$1\frac{1}{5} \times 2\frac{2}{3} \longrightarrow \frac{6}{5} \times \frac{8}{3} \longrightarrow \overset{2}{\cancel{6}}{5} \times \frac{8}{\cancel{3}}_{1} = \frac{16}{5} = 3\frac{1}{5}$$

$$4 \times 3\frac{3}{5} \longrightarrow \frac{4}{1} \times \frac{18}{5} \longrightarrow \frac{4}{1} \times \frac{18}{5} = \frac{72}{5} = 14\frac{2}{5}$$

Exercise A Multiply. Cancel when possible. Be sure each answer is in lowest terms.

1. $2\frac{1}{2} \times 4 =$ 2. $4\frac{1}{2} \times \frac{1}{3} =$

3. $2\frac{3}{4} \times 2\frac{2}{3} =$ 4. $4\frac{1}{5} \times \frac{3}{7} =$

5. $8 \times 2\frac{1}{12} =$ 6. $3 \times 4\frac{1}{9} =$

7. $2\frac{1}{2} \times 2\frac{1}{2} =$ 8. $1\frac{3}{10} \times 16 =$

9. $6\frac{1}{8} \times 4 =$ 10. $5 \times 1\frac{6}{7} =$

Exercise B Multiply. Write answers in lowest terms.

1. $3\frac{1}{3} \times 6 =$ 2. $5\frac{1}{4} \times \frac{1}{3} =$

3. $4\frac{1}{3} \times 3\frac{3}{4} =$ 4. $3\frac{1}{3} \times \frac{1}{5} =$

5. $3 \times 4\frac{1}{9} =$ 6. $7\frac{1}{3} \times \frac{1}{11} =$

7. $5 \times 5\frac{1}{11} =$ 8. $5\frac{3}{7} \times 1\frac{4}{9} =$

9. $3\frac{2}{3} \times 2\frac{1}{3} =$ 10. $1\frac{1}{5} \times 3\frac{2}{3} =$

11. $\frac{1}{16} \times 2\frac{2}{15} =$ 12. $4\frac{1}{4} \times 2\frac{1}{25} =$

LESSON EIGHT: Solving Word Problems

Directions   Read and solve each problem.

1. Sherrie needs $\frac{1}{3}$ of $\frac{1}{2}$ a pound of butter for a cake recipe. How much butter does she need?

   ----

2. Florentino has a $1\frac{1}{2}$-acre garden. He planted $\frac{1}{3}$ of it in carrots. How much land is planted in carrots?

   ----

3. It is suggested that no more than $\frac{1}{4}$ of a family's income go for rent. If a family makes $1,400 a month, what is the most that should go for rent?

   ----

4. As a general rule, mortgage companies suggest that a family can afford a house that costs no more than $2\frac{1}{2}$ times the family's total annual income. If you and your spouse have a combined income of $25,000 a year, can you afford a house that costs $70,000?

   ----

5. A recipe calls for $2\frac{1}{4}$ cups of flour, $1\frac{3}{4}$ quarts of milk, and $\frac{1}{2}$ pound of butter. If Calvin wants to make $2\frac{1}{2}$ times as much as the recipe yields, how much of each ingredient does he need?

   ----

6. Colleen knows that it takes $1\frac{1}{3}$ yards of material to make a dress for each of her daughters. If she plans to make Easter dresses for all three daughters, how many yards of material does she need?

   ----

7. You drive $5\frac{1}{2}$ miles to work each day, five days a week. How many miles do you drive per week to go to and from your job?

   ----

8. Meat should be cooked $\frac{1}{4}$ of an hour for each pound it weighs. For how long should you cook a 6-pound roast?

UNIT 6—REVIEW

Directions Multiply. Reduce each answer to lowest terms.

1. $5 \times \frac{1}{8} =$

2. $\frac{1}{4} \times 9 =$

3. $\frac{1}{4} \times 12 =$

4. $2\frac{1}{3} \times \frac{1}{3} =$

5. $\frac{3}{4} \times 1\frac{2}{5} =$

6. $4\frac{1}{2} \times 2\frac{2}{3} =$

7. $8\frac{1}{3} \times 1\frac{1}{5} =$

8. $\frac{1}{3} \times \frac{1}{3} =$

9. $\frac{9}{10} \times \frac{5}{12} =$

10. $4 \times \frac{1}{4} =$

11. $3 \times \frac{1}{5} =$

12. $\frac{1}{5} \times 6 =$

13. $\frac{1}{5} \times 25 =$

14. $3\frac{4}{5} \times \frac{1}{4} =$

15. $\frac{1}{2} \times 1\frac{1}{4} =$

16. $2\frac{2}{5} \times 6\frac{1}{2} =$

17. $2\frac{1}{2} \times 3\frac{1}{5} =$

18. $\frac{1}{5} \times \frac{1}{5} =$

19. $\frac{5}{8} \times \frac{4}{15} =$

20. $10 \times \frac{1}{10} =$

21. $\frac{1}{2} \times \frac{1}{2} =$

22. $1\frac{1}{2} \times 5\frac{4}{5} =$

23. $\frac{2}{3} \times \frac{3}{4} =$

24. $10 \times 3\frac{1}{2} =$

25. $4 \times \frac{3}{8} =$

26. $12\frac{2}{3} \times 5\frac{1}{3} =$

27. $9 \times \frac{1}{9} =$

28. $1\frac{1}{8} \times 3\frac{1}{3} =$

29. $2\frac{2}{3} \times 7 =$

30. $4\frac{2}{3} \times 4\frac{1}{2} =$

31. $18 \times \frac{1}{4} =$

32. $2\frac{2}{3} \times 5\frac{1}{3} =$

33. $\frac{7}{8} \times \frac{4}{21} =$

34. $12 \times \frac{3}{4} =$

COMPARISON SHOPPING

When buying meat, it is helpful to consider the cost per pound and the cost per serving. Some meats contain bone, gristle, and fat waste. If you know the cost per pound of a cut of meat and how many servings it will make, you can calculate the cost per person. The chart below shows the yield per person, the average cost per pound, and the cost per serving for various cuts of meat.

Cut of Beef	Allow Per Person (one serving)	Cost Per Pound	Cost Per Serving
rib roast	$\frac{7}{8}$ lb.	$2.36	$2.07
sirloin tip roast	$\frac{1}{3}$ lb.	$2.86	$ .95
rump roast, boneless	$\frac{1}{3}$ lb.	$2.49	$ .83
rump roast, bone-in	$\frac{1}{2}$ lb.	$2.11	$1.06
chicken wing	$1\frac{1}{2}$ lb.	$ .61	$ .92
ham, fully-cooked boneless	$\frac{1}{4}$ lb.	$3.20	$ .80
arm roast (pot roast)	$\frac{1}{2}$ lb.	$1.86	$ .93
club steak	$\frac{7}{8}$ lb.	$2.49	$2.18
sirloin strip steak	$\frac{5}{8}$ lb.	$2.61	$1.63
rib-eye steak	$\frac{5}{8}$ lb.	$4.49	$2.81

Directions Using the information in the chart, calculate the amount of meat needed to feed each group. The first one has been done.

	Number of People	Type of Meat	Pounds Needed
	4	boneless rump roast	$4 \times \frac{1}{3} = 1\frac{1}{3}$
1.	10	chicken wing	
2.	5	fully-cooked, boneless ham	
3.	2	rib-eye steak	
4.	12	pot roast	
5.	8	club steak	
6.	9	rump roast, bone-in	
7.	12	sirloin tip roast	

UNIT 7—PRETEST
DIVIDING FRACTIONS

Divide these fractions. Blacken the letter to the right that corresponds to the correct answer.

1. $\frac{1}{3} \div \frac{3}{8} =$ a. $\frac{1}{8}$ b. $\frac{8}{9}$ c. $\frac{4}{11}$ d. $\frac{3}{24}$ a b c d

2. $\frac{2}{11} \div \frac{1}{2} =$ a. $\frac{4}{11}$ b. $\frac{1}{9}$ c. $\frac{2}{22}$ d. $\frac{2}{13}$ a b c d

3. $\frac{1}{5} \div \frac{5}{6} =$ a. $\frac{6}{11}$ b. $\frac{4}{11}$ c. $\frac{6}{25}$ d. $\frac{1}{6}$ a b c d

4. $\frac{1}{8} \div \frac{2}{7} =$ a. $\frac{1}{5}$ b. $\frac{7}{16}$ c. $\frac{2}{56}$ d. $\frac{1}{28}$ a b c d

5. $\frac{1}{4} \div \frac{1}{2} =$ a. 4   b. $\frac{1}{3}$ c. $\frac{1}{8}$ d. $\frac{1}{2}$ a b c d

6. $\frac{2}{3} \div \frac{5}{6} =$ a. $\frac{4}{5}$ b. $\frac{5}{9}$ c. $\frac{7}{9}$ d. $\frac{1}{5}$ a b c d

7. $\frac{1}{16} \div \frac{1}{8} =$ a. $\frac{1}{2}$ b. $\frac{1}{12}$ c. $\frac{1}{8}$ d. $\frac{1}{4}$ a b c d

8. $\frac{2}{3} \div \frac{8}{9} =$ a. $1\frac{1}{3}$ b. $\frac{3}{4}$ c. 1   d. $\frac{5}{6}$ a b c d

9. $\frac{1}{2} \div \frac{1}{4} =$ a. $\frac{1}{8}$ b. $\frac{2}{8}$ c. 8   d. 2   a b c d

10. $\frac{3}{8} \div \frac{1}{3} =$ a. $\frac{2}{5}$ b. $\frac{1}{8}$ c. $1\frac{1}{8}$ d. $\frac{8}{9}$ a b c d

11. $\frac{5}{6} \div \frac{1}{3} =$ a. $\frac{5}{18}$ b. $1\frac{1}{3}$ c. $2\frac{1}{2}$ d. $\frac{4}{3}$ a b c d

12. $\frac{5}{9} \div \frac{1}{6} =$ a. $\frac{3}{10}$ b. $\frac{5}{54}$ c. $1\frac{1}{3}$ d. $3\frac{1}{3}$ a b c d

13. $5 \div \frac{1}{4} =$ a. 20   b. $1\frac{1}{4}$ c. $\frac{1}{4}$ d. 54   a b c d

14. $16 \div \frac{4}{7} =$ a. $9\frac{1}{7}$ b. 28   c. $16\frac{4}{7}$ d. $\frac{4}{7}$ ⬚a ⬚b ⬚c ⬚d

15. $10 \div \frac{2}{3} =$ a. $\frac{20}{3}$ b. $6\frac{2}{3}$ c. 15   d. $\frac{1}{15}$ ⬚a ⬚b ⬚c ⬚d

16. $27 \div \frac{9}{11} =$ a. 33   b. $27\frac{9}{11}$ c. 18   d. $\frac{3}{11}$ ⬚a ⬚b ⬚c ⬚d

17. $\frac{1}{3} \div 4 =$ a. $4\frac{1}{3}$ b. $1\frac{1}{3}$ c. $\frac{1}{12}$ d. 12   ⬚a ⬚b ⬚c ⬚d

18. $\frac{3}{7} \div 6 =$ a. $2\frac{4}{7}$ b. $5\frac{4}{7}$ c. $6\frac{3}{7}$ d. $\frac{1}{14}$ ⬚a ⬚b ⬚c ⬚d

19. $\frac{4}{5} \div 24 =$ a. $23\frac{1}{5}$ b. $\frac{1}{30}$ c. $24\frac{4}{5}$ d. $\frac{1}{24}$ ⬚a ⬚b ⬚c ⬚d

20. $\frac{7}{9} \div 21 =$ a. $\frac{1}{27}$ b. $\frac{1}{3}$ c. $21\frac{7}{9}$ d. $20\frac{2}{9}$ ⬚a ⬚b ⬚c ⬚d

21. $1\frac{1}{3} \div 2\frac{1}{5} =$ a. $3\frac{2}{15}$ b. $3\frac{8}{15}$ c. $\frac{7}{15}$ d. $\frac{20}{33}$ ⬚a ⬚b ⬚c ⬚d

22. $2\frac{1}{8} \div 2\frac{1}{4} =$ a. $\frac{27}{43}$ b. $4\frac{3}{8}$ c. $\frac{1}{4}$ d. $\frac{17}{18}$ ⬚a ⬚b ⬚c ⬚d

23. $3\frac{1}{3} \div 3\frac{1}{3} =$ a. 0   b. 1   c. $6\frac{2}{3}$ d. 7   ⬚a ⬚b ⬚c ⬚d

24. $2\frac{1}{9} \div 7\frac{1}{3} =$ a. $\frac{19}{66}$ b. $9\frac{4}{9}$ c. $5\frac{2}{9}$ d. $\frac{22}{27}$ ⬚a ⬚b ⬚c ⬚d

Read and solve each problem. Blacken the letter to the right that corresponds to the correct answer.

25. Martina has $12\frac{2}{5}$ pounds of clay to use in 5 projects. How many pounds of clay can be used for each project?

a $7\frac{2}{5}$ b $\frac{5}{62}$ c $2\frac{12}{25}$ d $2\frac{1}{5}$

[a] [b] [c] [d]

26. It usually takes Germaine and Tony $2\frac{3}{4}$ hours to prepare the foundation for an above-ground pool. How many foundations can they prepare in 44 hours of work?

a. 16   b. $20\frac{3}{4}$ c. 121   d. $11\frac{4}{7}$

[a] [b] [c] [d]

27. The Millers have 426 acres of farmland. If they keep $2\frac{1}{2}$ acres for their home and divide the rest equally among their 5 children, how many acres will each child receive?

a. $85\frac{7}{10}$ b. $80\frac{2}{15}$ c. $85\frac{1}{5}$ d. $84\frac{7}{10}$

[a] [b] [c] [d]

28. Delia Moros, an interviewer with an employment agency, schedules appointments $\frac{3}{4}$ of an hour apart. How many interviews can she conduct in 6 hours?

a. 6   b. 8   c. $6\frac{3}{4}$ d. $5\frac{1}{4}$

[a] [b] [c] [d]

29. Mr. Cardenas bought a 100-acre tract for a subdivision. He figures that $\frac{1}{10}$ of the land will be used for streets. How many $\frac{3}{8}$-acre lots can the remainder of the land be divided into?

a. 240   b. $226\frac{2}{3}$ c. 100   d. $100\frac{1}{10}$

[a] [b] [c] [d]

30. Fred filled his car before starting on a trip. He then drove $265\frac{5}{10}$ miles. When he arrived at his destination, he filled his gas tank again. Fred next figured his gas mileage by dividing the amount of gas used ($17\frac{7}{10}$ gallons) into the number of miles traveled. How many miles per gallon did Fred's car average on the trip?

a. $250\frac{6}{10}$ b. 15   c. $265\frac{1}{2}$ d. 23

[a] [b] [c] [d]

DIVIDING FRACTIONS

Dividing Fractions

Instruction
To divide fractions, first turn the right-hand fraction upside down (**invert**). Then multiply the numerator times the numerator and the denominator times the denominator.

Example

$$\frac{1}{4} \div \frac{1}{3} \longrightarrow \frac{1}{4} \times \frac{3}{1} \longrightarrow \frac{1}{4} \times \frac{3}{1} = \frac{3}{4}$$

$$\frac{2}{3} \div \frac{5}{7} \longrightarrow \frac{2}{3} \times \frac{7}{5} \longrightarrow \frac{2}{3} \times \frac{7}{5} = \frac{14}{15}$$

Exercise A   Divide.

1. $\frac{1}{2} \div \frac{2}{3} =$

2. $\frac{1}{4} \div \frac{1}{3} =$

3. $\frac{1}{3} \div \frac{3}{4} =$

4. $\frac{1}{5} \div \frac{2}{3} =$

5. $\frac{1}{2} \div \frac{7}{9} =$

6. $\frac{3}{5} \div \frac{2}{3} =$

7. $\frac{1}{6} \div \frac{1}{5} =$

8. $\frac{2}{3} \div \frac{1}{7} =$

9. $\frac{3}{7} \div \frac{1}{2} =$

10. $\frac{4}{11} \div \frac{1}{3} =$

11. $\frac{7}{16} \div \frac{1}{5} =$

12. $\frac{3}{7} \div \frac{2}{3} =$

Exercise B   Divide.

1. $\frac{1}{3} \div \frac{3}{8} =$

2. $\frac{1}{4} \div \frac{2}{5} =$

3. $\frac{2}{5} \div \frac{5}{8} =$

4. $\frac{3}{8} \div \frac{4}{5} =$

5. $\frac{1}{3} \div \frac{1}{2} =$

6. $\frac{3}{5} \div \frac{7}{11} =$

7. $\frac{2}{3} \div \frac{3}{8} =$

8. $\frac{3}{10} \div \frac{1}{2} =$

9. $\frac{1}{7} \div \frac{1}{2} =$

10. $\frac{1}{11} \div \frac{2}{5} =$

LESSON TWO: Writing Division Problems

Instruction It is important to be able to read a division problem. In division, one number is divided by another number to obtain the answer.

Example

Long Division	Fractional Division
$\dfrac{\text{Quotient}}{\text{Divisor} \,)\, \text{Dividend}}$	Dividend ÷ Divisor = Quotient
$2\overline{)6}^{\;3}$	$\frac{1}{4} \div \frac{1}{3} =$

This sign (÷) means divided by.

Example

$\frac{1}{4} \div \frac{1}{3}$ means $\frac{1}{4}$ divided by $\frac{1}{3}$, or $\frac{1}{3}$ divided into $\frac{1}{4}$.

$\frac{2}{5} \div \frac{1}{5}$ means $\frac{2}{5}$ divided by $\frac{1}{5}$, or $\frac{1}{5}$ divided into $\frac{2}{5}$.

$\frac{1}{7} \div \frac{2}{3}$ means $\frac{1}{7}$ divided by $\frac{2}{3}$, or $\frac{2}{3}$ divided into $\frac{1}{7}$.

$\frac{3}{8} \div \frac{1}{7}$ means $\frac{3}{8}$ divided by $\frac{1}{7}$, or $\frac{1}{7}$ divided into $\frac{3}{8}$.

Exercise A Write these expressions as division problems. Do not solve the problems.

1. $\frac{1}{2}$ divided by $\frac{3}{4}$ _____

2. $\frac{1}{2}$ divided into $\frac{3}{5}$ _____

3. $\frac{2}{3}$ divided by $\frac{1}{4}$ _____

4. $\frac{2}{5}$ divided into $\frac{2}{5}$ _____

5. divide $\frac{1}{8}$ into $\frac{4}{7}$ _____

6. $\frac{7}{8}$ divided by $\frac{3}{4}$ _____

7. $\frac{5}{6}$ divided into $\frac{1}{2}$ _____

8. $\frac{1}{9}$ divided into $\frac{1}{3}$ _____

9. $\frac{2}{7}$ divided by $\frac{1}{4}$ _____

10. $\frac{1}{6}$ divided by $\frac{1}{2}$ _____

Exercise B Write these expressions as division problems. Do not solve them.

1. $\frac{2}{3}$ divided by $\frac{3}{8}$ _____

2. divide $\frac{1}{2}$ into $\frac{3}{4}$ _____

3. $\frac{2}{3}$ divided into $\frac{7}{8}$ _____

4. $\frac{3}{5}$ divided by $\frac{1}{4}$ _____

5. divide $\frac{3}{5}$ into $\frac{4}{7}$ _____

6. $\frac{2}{7}$ divided by $\frac{3}{4}$ _____

7. $\frac{1}{4}$ divided by $\frac{3}{11}$ _____

8. $\frac{3}{8}$ divided into $\frac{2}{9}$ _____

9. $\frac{2}{5}$ divided by $\frac{1}{5}$ _____

10. $\frac{2}{3}$ divided by $\frac{1}{6}$ _____

Cancelling in Division

Instruction

In division, after inverting the divisor (the right hand fraction), you may cancel when possible before multiplying numerators and denominators.

Example

$$\frac{1}{2} \div \frac{3}{4} \longrightarrow \frac{1}{2} \times \frac{4}{3} \longrightarrow \frac{1}{\overset{}{\underset{1}{2}}} \times \frac{\overset{2}{4}}{3} = \frac{2}{3}$$

$$\frac{3}{7} \div \frac{9}{10} \longrightarrow \frac{3}{7} \times \frac{10}{9} \longrightarrow \frac{\overset{1}{3}}{7} \times \frac{10}{\underset{3}{9}} = \frac{10}{21}$$

$$\frac{5}{8} \div \frac{5}{6} \longrightarrow \frac{5}{8} \times \frac{6}{5} \longrightarrow \frac{\overset{1}{5}}{\underset{4}{8}} \times \frac{\overset{3}{6}}{\underset{1}{5}} = \frac{3}{4}$$

Remember, sometimes you cannot cancel any numbers. Sometimes you can cancel two numbers. Sometimes you can cancel all numbers.

$$\frac{1}{6} \div \frac{7}{8} \longrightarrow \frac{1}{6} \times \frac{8}{7} \longrightarrow \frac{1}{\underset{3}{6}} \times \frac{\overset{4}{8}}{7} = \frac{4}{21}$$

$$\frac{3}{7} \div \frac{4}{5} \longrightarrow \frac{3}{7} \times \frac{5}{4} \longrightarrow \frac{3}{7} \times \frac{5}{4} = \frac{15}{28}$$

$$\frac{3}{16} \div \frac{3}{8} \longrightarrow \frac{3}{16} \times \frac{8}{3} \longrightarrow \frac{\overset{1}{3}}{\underset{2}{16}} \times \frac{\overset{1}{8}}{\underset{1}{3}} = \frac{1}{2}$$

Exercise A

Divide. Cancel when possible.

1. $\frac{1}{2} \div \frac{3}{4} =$ 　　　　　　　　2. $\frac{1}{6} \div \frac{1}{3} =$

3. $\frac{1}{2} \div \frac{7}{8} =$ 　　　　　　　　4. $\frac{2}{3} \div \frac{5}{6} =$

5. $\frac{1}{4} \div \frac{5}{5} =$ 　　　　　　　　6. $\frac{1}{8} \div \frac{1}{4} =$

7. $\frac{1}{8} \div \frac{3}{10} =$ 　　　　　　　　8. $\frac{3}{8} \div \frac{1}{2} =$

9. $\frac{1}{8} \div \frac{7}{8} =$ 10. $\frac{3}{7} \div \frac{4}{7} =$

11. $\frac{1}{5} \div \frac{4}{4} =$ 12. $\frac{1}{6} \div \frac{5}{6} =$

13. $\frac{3}{7} \div \frac{9}{10} =$ 14. $\frac{1}{6} \div \frac{7}{8} =$

Exercise B   Divide. Cancel when possible.

1. $\frac{3}{4} \div \frac{3}{4} =$ 2. $\frac{2}{3} \div \frac{2}{3} =$

3. $\frac{1}{3} \div \frac{3}{5} =$ 4. $\frac{1}{4} \div \frac{4}{5} =$

5. $\frac{5}{8} \div \frac{5}{6} =$ 6. $\frac{1}{4} \div \frac{1}{4} =$

7. $\frac{1}{4} \div \frac{3}{8} =$ 8. $\frac{3}{16} \div \frac{3}{8} =$

9. $\frac{2}{7} \div \frac{2}{3} =$ 10. $\frac{4}{9} \div \frac{4}{9} =$

11. $\frac{1}{6} \div \frac{5}{9} =$ 12. $\frac{3}{16} \div \frac{3}{4} =$

13. $\frac{1}{7} \div \frac{3}{7} =$ 14. $\frac{1}{7} \div \frac{4}{7} =$

LESSON FOUR: Simplifying Answers

Instruction

Every answer should be reduced to lowest terms. Remember, a proper fraction is reduced by dividing the numerator and the denominator by the same number. An improper fraction is reduced by dividing the numerator by the denominator.

Example

$$\frac{1}{4} \div \frac{3}{4} \longrightarrow \frac{1}{4} \times \frac{4}{3} = \frac{4}{12} = \frac{1}{3}$$

$$\frac{3}{4} \div \frac{1}{2} \longrightarrow \frac{3}{4} \times \frac{2}{1} = \frac{6}{4} = 1\frac{2}{4} = 1\frac{1}{2}$$

$$\frac{1}{2} \div \frac{1}{2} \longrightarrow \frac{1}{2} \times \frac{2}{1} = \frac{2}{2} = 1$$

Exercise A

Divide. Reduce each answer to its lowest term if necessary.

1. $\frac{2}{3} \div \frac{5}{8} =$

2. $\frac{3}{4} \div \frac{2}{5} =$

3. $\frac{3}{4} \div \frac{2}{3} =$

4. $\frac{1}{2} \div \frac{2}{3} =$

5. $\frac{7}{8} \div \frac{1}{2} =$

6. $\frac{4}{7} \div \frac{2}{3} =$

7. $\frac{2}{7} \div \frac{1}{3} =$

8. $\frac{3}{11} \div \frac{4}{5} =$

9. $\frac{5}{6} \div \frac{2}{3} =$

10. $\frac{2}{3} \div \frac{1}{6} =$

11. $\frac{5}{7} \div \frac{2}{7} =$

12. $\frac{2}{5} \div \frac{1}{5} =$

13. $\frac{3}{8} \div \frac{1}{4} =$

14. $\frac{3}{5} \div \frac{1}{4} =$

15. $\frac{2}{7} \div \frac{3}{14} =$

16. $\frac{1}{2} \div \frac{2}{7} =$

17. $\frac{5}{8} \div \frac{1}{2} =$

18. $\frac{4}{4} \div \frac{3}{7} =$

19. $\frac{7}{8} \div \frac{1}{6} =$

20. $\frac{3}{4} \div \frac{1}{2} =$

21. $\frac{5}{6} \div \frac{5}{8} =$ 22. $\frac{4}{5} \div \frac{1}{5} =$

23. $\frac{4}{7} \div \frac{1}{5} =$ 24. $\frac{5}{7} \div \frac{3}{5} =$

25. $\frac{3}{7} \div \frac{1}{3} =$ 26. $\frac{1}{16} \div \frac{1}{20} =$

27. $\frac{4}{5} \div \frac{1}{11} =$ 28. $\frac{3}{14} \div \frac{3}{16} =$

Exercise B

Divide. Reduce answers to lowest terms if necessary.

1. $\frac{1}{4} \div \frac{1}{4} =$ 2. $\frac{3}{5} \div \frac{3}{5} =$

3. $\frac{4}{5} \div \frac{1}{2} =$ 4. $\frac{2}{5} \div \frac{1}{4} =$

5. $\frac{7}{8} \div \frac{1}{4} =$ 6. $\frac{2}{7} \div \frac{4}{7} =$

7. $\frac{2}{3} \div \frac{4}{9} =$ 8. $\frac{2}{5} \div \frac{3}{5} =$

9. $\frac{5}{9} \div \frac{1}{6} =$ 10. $\frac{2}{9} \div \frac{1}{12} =$

11. $\frac{1}{5} \div \frac{1}{8} =$ 12. $\frac{3}{4} \div \frac{1}{3} =$

13. $\frac{5}{12} \div \frac{1}{3} =$ 14. $\frac{2}{5} \div \frac{1}{8} =$

15. $\frac{5}{7} \div \frac{1}{6} =$ 16. $\frac{5}{6} \div \frac{3}{8} =$

17. $\frac{7}{9} \div \frac{1}{15} =$ 18. $\frac{1}{10} \div \frac{1}{15} =$

19. $\frac{8}{11} \div \frac{1}{3} =$ 20. $\frac{7}{10} \div \frac{1}{9} =$

21. $\frac{5}{11} \div \frac{1}{6} =$ 22. $\frac{7}{12} \div \frac{1}{4} =$

23. $\frac{8}{9} \div \frac{1}{3} =$ 24. $\frac{1}{10} \div \frac{1}{3} =$

Writing Whole Numbers in Fractional Form

Instruction To write a whole number as an improper fraction, place the whole number over 1.

Example

$$6 = \frac{6}{1} \qquad 12 = \frac{12}{1} \qquad 1 = \frac{1}{1}$$

Exercise A Write the following whole numbers in fractional form.

1. 6 =   2. 23 =   3. 18 =

4. 100 =   5. 5 =   6. 50 =

7. 10 =   8. 2 =   9. 1 =

10. 7 =   11. 101 =   12. 44 =

13. 2 =   14. 51 =   15. 27 =

16. 24 =   17. 23 =   18. 10 =

Exercise B Write these whole numbers in fractional form.

1. 11 =   2. 20 =   3. 45 =

4. 3 =   5. 19 =   6. 100 =

7. 21 =   8. 36 =   9. 145 =

10. 110 =   11. 135 =   12. 119 =

13. 39 =   14. 77 =   15. 4 =

16. 14 =   17. 1 =   18. 5 =

19. 38 =   20. 98 =   21. 16 =

22. 76 =   23. 40 =   24. 84 =

LESSON SIX:

Dividing With a Whole Number in the Dividend

Instruction

When the dividend is a whole number, write the whole number in fractional form by placing it over 1. Then invert and multiply.

Example

$$3 \div \frac{1}{2} \longrightarrow \frac{3}{1} \div \frac{1}{2} \longrightarrow \frac{3}{1} \times \frac{2}{1} = \frac{6}{1} = 6$$

$$4 \div \frac{2}{3} \longrightarrow \frac{4}{1} \div \frac{2}{3} \longrightarrow \frac{\overset{2}{\cancel{4}}}{1} \times \frac{3}{\underset{1}{\cancel{2}}} = \frac{6}{1} = 6$$

Exercise A   Divide and simplify.

1. $3 \div \frac{2}{3} =$ 2. $5 \div \frac{2}{5} =$

3. $15 \div \frac{1}{3} =$ 4. $21 \div \frac{1}{3} =$

5. $6 \div \frac{1}{6} =$ 6. $4 \div \frac{1}{4} =$

7. $23 \div \frac{1}{3} =$ 8. $39 \div \frac{1}{2} =$

9. $11 \div \frac{1}{2} =$ 10. $18 \div \frac{3}{7} =$

Exercise B   Divide and simplify.

1. $4 \div \frac{2}{3} =$ 2. $2 \div \frac{3}{5} =$

3. $20 \div \frac{3}{8} =$ 4. $24 \div \frac{6}{7} =$

5. $10 \div \frac{2}{5} =$ 6. $45 \div \frac{1}{3} =$

7. $28 \div \frac{3}{4} =$ 8. $18 \div \frac{1}{6} =$

9. $4 \div \frac{2}{7} =$ 10. $38 \div \frac{2}{3} =$

Dividing With a Whole Number as the Divisor

Instruction

When the divisor is a whole number, first write it as an improper fraction. Then invert and multiply.

Example

$$\frac{2}{3} \div 3 \longrightarrow \frac{2}{3} \div \frac{3}{1} \longrightarrow \frac{2}{3} \times \frac{1}{3} = \frac{2}{9}$$

$$\frac{1}{5} \div 5 \longrightarrow \frac{1}{5} \div \frac{5}{1} \longrightarrow \frac{1}{5} \times \frac{1}{5} = \frac{1}{25}$$

Remember, a whole number is equal to itself over the number 1. $4 = \frac{4}{1}$. Also remember that the divisor (the right hand fraction) must be inverted before multiplying. 4 equals $\frac{4}{1}$. However, 4 **inverted** is $\frac{1}{4}$.

Exercise A

Divide. Write answers in lowest terms.

1. $\frac{2}{9} \div 3 =$ 2. $\frac{3}{5} \div 5 =$

3. $\frac{1}{8} \div 5 =$ 4. $\frac{1}{7} \div 11 =$

5. $\frac{2}{3} \div 2 =$ 6. $\frac{4}{5} \div 4 =$

7. $\frac{3}{8} \div 12 =$ 8. $\frac{4}{7} \div 4 =$

9. $\frac{4}{5} \div 6 =$ 10. $\frac{2}{7} \div 14 =$

Exercise B

Divide. Write answers in lowest terms.

1. $\frac{5}{6} \div 10 =$ 2. $\frac{5}{8} \div 15 =$

3. $\frac{3}{4} \div 12 =$ 4. $\frac{5}{7} \div 25 =$

5. $\frac{2}{7} \div 6 =$ 6. $\frac{3}{4} \div 9 =$

Dividing With Mixed Numbers

Instruction In division, whenever there is a mixed number, first change it to an improper fraction. Then invert the divisor and multiply.

Example

$$1\frac{1}{2} \div 2\frac{2}{3} \longrightarrow \frac{3}{2} \div \frac{8}{3} \longrightarrow \frac{3}{2} \times \frac{3}{8} = \frac{9}{16}$$

$$\frac{2}{5} \div 1\frac{1}{4} \longrightarrow \frac{2}{5} \div \frac{5}{4} \longrightarrow \frac{2}{5} \times \frac{4}{5} = \frac{8}{25}$$

$$6 \div 3\frac{3}{4} \longrightarrow \frac{6}{1} \div \frac{15}{4} \longrightarrow \frac{\overset{2}{6}}{1} \times \frac{4}{\underset{5}{15}} = \frac{8}{5} = 1\frac{3}{5}$$

Exercise A Divide.

1. $2\frac{2}{3} \div 2\frac{1}{2} =$

2. $2 \div 3\frac{2}{3} =$

3. $2\frac{1}{3} \div 1\frac{4}{5} =$

4. $2\frac{1}{2} \div \frac{2}{3} =$

5. $1\frac{1}{4} \div 2\frac{2}{5} =$

6. $\frac{2}{3} \div 1\frac{1}{5} =$

7. $\frac{3}{4} \div 3\frac{1}{2} =$

8. $4 \div 1\frac{5}{8} =$

9. $\frac{1}{2} \div 3\frac{1}{4} =$

10. $2\frac{1}{2} \div 7\frac{1}{2} =$

Exercise B Divide.

1. $2\frac{1}{3} \div 2\frac{2}{3} =$

2. $2 \div 1\frac{1}{2} =$

3. $2\frac{2}{7} \div 2\frac{1}{3} =$

4. $1\frac{3}{4} \div \frac{3}{4} =$

5. $2\frac{1}{5} \div 2\frac{1}{3} =$

6. $\frac{3}{5} \div 4\frac{1}{2} =$

LESSON NINE: Solving Word Problems

Directions   Read and solve each problem.

1. The trotline Mr. Robinson is putting in the river will be 20 yards long. If he wants to have a hook every $\frac{1}{3}$ of a yard, how many hooks will he need?

2. Esmond bought a 12-pound case of formula for his baby. The formula was in $\frac{1}{2}$-pound cans. How many cans were there in the case?

3. A bricklayer has to build 6 walls with $4\frac{2}{5}$ tons of bricks. How many tons of bricks will there be in each wall?

4. Lester tries to study his homework $1\frac{1}{2}$ hours each night. He has 4 classes to study for. How much time can he spend on each class if he devotes the same amount of time to each one?

5. Florencio's actual work day is $7\frac{1}{2}$ hours long. If it is divided into six work periods, how long is each period?

6. Joan can usually mow a small yard in $\frac{3}{4}$ of an hour. How many yards can she mow in 6 hours?

7. Mrs. Vickery bought 15 yards of material to make towels. Each towel requires $\frac{3}{4}$ yard of material. How many towels can she make from the material?

8. Allison cooked a $7\frac{1}{2}$-pound roast for the dinner. There will be 10 people for dinner. How much roast may each person have, assuming there is a two-pound weight loss during cooking?

92

UNIT 7—REVIEW

Divide. Write each answer in lowest terms.

1. $\frac{1}{3} \div \frac{1}{6} =$

2. $\frac{3}{5} \div 6 =$

3. $\frac{3}{5} \div \frac{9}{10} =$

4. $5 \div 2\frac{1}{2} =$

5. $6 \div \frac{1}{2} =$

6. $8\frac{1}{3} \div 1\frac{1}{4} =$

7. $15 \div \frac{3}{5} =$

8. $22\frac{1}{2} \div 1\frac{4}{5} =$

9. $\frac{5}{8} \div 10 =$

10. $3\frac{1}{3} \div 10 =$

11. $7 \div \frac{1}{3} =$

12. $\frac{2}{3} \div \frac{1}{5} =$

13. $1\frac{1}{2} \div 6 =$

14. $3\frac{2}{3} \div 1\frac{4}{5} =$

15. $2 \div \frac{1}{4} =$

16. $13 \div \frac{1}{2} =$

17. $8\frac{1}{3} \div 12 =$

18. $\frac{2}{5} \div \frac{7}{8} =$

19. $5 \div 20 =$

20. $1\frac{1}{4} \div 2\frac{1}{2} =$

21. $\frac{2}{7} \div \frac{1}{5} =$

22. $24 \div 7\frac{2}{5} =$

23. $6\frac{3}{8} \div \frac{5}{6} =$

24. $\frac{9}{10} \div 7 =$

25. $7 \div \frac{1}{3} =$

26. $14\frac{5}{12} \div \frac{1}{2} =$

27. $11 \div \frac{3}{8} =$

28. $16\frac{1}{8} \div \frac{1}{3} =$

29. $\frac{3}{4} \div 24 =$

30. $17\frac{1}{2} \div 17 =$

Directions

Divide. Write answers in lowest terms.

1. $\frac{1}{6} \div \frac{1}{3} =$

2. $\frac{1}{2} \div 3 =$

3. $\frac{3}{8} \div \frac{3}{5} =$

4. $11 \div 5\frac{1}{2} =$

5. $10 \div \frac{2}{3} =$

6. $6\frac{2}{5} \div 5\frac{1}{3} =$

7. $14 \div \frac{2}{7} =$

8. $17\frac{1}{2} \div 3\frac{1}{2} =$

9. $\frac{4}{5} \div 8 =$

10. $3\frac{1}{3} \div 5 =$

11. $\frac{3}{4} \div \frac{1}{2} =$

12. $\frac{8}{25} \div \frac{4}{5} =$

13. $2\frac{1}{3} \div 1\frac{2}{3} =$

14. $\frac{3}{4} \div 12 =$

15. $\frac{1}{10} \div \frac{1}{2} =$

16. $10\frac{1}{2} \div 1\frac{2}{5} =$

17. $5 \div \frac{2}{5} =$

18. $\frac{5}{8} \div 1\frac{1}{4} =$

19. $\frac{3}{8} \div 3 =$

20. $8 \div 1\frac{1}{2} =$

21. A community has a $4\frac{1}{2}$-acre lot which is to be divided into $\frac{1}{2}$-acre gardens. Each garden spot will be rented to one family. How many families can have a $\frac{1}{2}$-acre garden spot? -------------------

22. Together, 4 boys pick up $10\frac{1}{5}$ pounds of aluminum cans. How many pounds will each get if they divide the cans equally before taking them to the recycling center? -------------------

Reynolds Aluminum

94

COMPARISON SHOPPING

Wise grocery shoppers compare prices. But, you cannot always know the better buy unless you also consider how much of the item you will be getting. You much look at the net contents and price of the items you are comparing, not just the size of the container. There are many odd-shaped containers. The amount inside can fool you.

To compare prices, it is helpful to convert the total price of similar items to **unit prices**. The unit price is how much an item costs per pound, ounce, pint, etc. **To find the cost per unit, divide the cost of any item by the amount in the container.** Some stores have unit prices already marked on their packages. They have done the figuring for you.

Directions

Look at the following chart. Circle the item in each pair that is the better buy.

Item	Quantity	Price	Unit Price
beans, brand X	16 oz.	$ .43	$2\frac{7}{10}$¢ per oz.
beans, brand Y	12 oz.	$ .37	$3\frac{1}{10}$¢ per oz.
detergent, regular size	$5\frac{1}{4}$ lb.	$2.69	$56\frac{8}{10}$¢ per lb.
detergent, king size	$10\frac{1}{4}$ lb.	$5.49	$53\frac{6}{10}$¢ per lb.
cheese pizza	$13\frac{1}{2}$ oz.	$1.42	$10\frac{5}{10}$¢ per oz.
sausage pizza	$13\frac{1}{2}$ oz.	$1.74	$12\frac{9}{10}$¢ per oz.
frozen roast dinner	$6\frac{1}{2}$ oz.	$2.49	$38\frac{3}{10}$¢ per oz.
frozen turkey dinner	$11\frac{1}{2}$ oz.	$2.01	$17\frac{5}{10}$¢ per oz.

Directions

Find the cost per unit of several items you buy regularly. Use the space below to make a chart of your findings.

FINAL REVIEW

Read the problems which follow. Compute the answer to each problem and blacken the letter to the right that corresponds to the correct answer.

1. 9,806 + 27,553 + 22 + 806,533 =   a. 897,464   ⬚a ⬚b ⬚c ⬚d
   b. 897,904   c. 843,914   d. 843,904

2. 2 + 19 + 1,304 + 7 =
   a. 1,332   b. 2,132   c. 1,732   d. 1,722   ⬚a ⬚b ⬚c ⬚d

3.   7,233   a. 17,224   b. 17,204   c. 17,324   ⬚a ⬚b ⬚c ⬚d
       415   d. 16,224
       671
   + 8,905

4.   25,531   a. 48,423   b. 47,413   c. 46,413   ⬚a ⬚b ⬚c ⬚d
      9,065   d. 48,403
     10,031
       766
   + 3,030

5. 89,628 − 367 = a. 88,261   b. 89,361   c. 89,995   ⬚a ⬚b ⬚c ⬚d
                  d. 89,261

6. 301 − 29 = a. 330   b. 282   c. 320   d. 272   ⬚a ⬚b ⬚c ⬚d

7.   9,362   a. 8,989   b. 9,735   c. 9,011   d. 9,099   ⬚a ⬚b ⬚c ⬚d
    − 373

8.   625,053   a. 585,239   b. 585,229   c. 664,877   ⬚a ⬚b ⬚c ⬚d
    − 39,824   d. 685,239

9. 62,479 × 8 =   a. 489,732   b. 479,622   c. 499,832   ⬚a ⬚b ⬚c ⬚d
                    d. 498,832

10. 3,003 × 9 = a. 29,007   b. 3,012   c. 27,027   ⬚a ⬚b ⬚c ⬚d
                d. 2,727

11.   47,821   a. 95,642   b. 956,420   c. 9,564,200   ⬚a ⬚b ⬚c ⬚d
    × 2,000   d. 95,642,000

12. 293,536
    × 706

a. 207,236,416   b. 27,236   c. 207,416
d. 207,236

☐a ☐b ☐c ☐d

13. 7) 805

a. 915   b. 115   c. 105   d. 195

☐a ☐b ☐c ☐d

14. 3) 7,634

a. 2,645   b. 2,544 r2   c. 2,211 r1
d. 2,508 r2

☐a ☐b ☐c ☐d

15. 20) 11,307

a. 565 r7   b. 575   c. 565
d. 564 r20

☐a ☐b ☐c ☐d

16. 14) 67,678

a. 4,834   b. 6,424 r2   c. 4,834 r2
d. 4,833 r16

☐a ☐b ☐c ☐d

17. $\frac{1}{9} + \frac{1}{9} + \frac{1}{9} + \frac{1}{9} + \frac{1}{9} + \frac{1}{9} =$
    a. $\frac{1}{54}$ b. $\frac{7}{9}$ c. $\frac{8}{9}$ d. $\frac{2}{3}$

☐a ☐b ☐c ☐d

18. $\frac{3}{11} + \frac{7}{11} + \frac{5}{11} =$ a. $1\frac{4}{11}$ b. $\frac{15}{33}$ c. $\frac{15}{22}$ d. $2\frac{3}{15}$

☐a ☐b ☐c ☐d

19. $\frac{7}{12}$

  $+ \frac{1}{3}$

a. $\frac{8}{15}$ b. $\frac{11}{15}$ c. $\frac{11}{12}$ d. $1\frac{1}{12}$

☐a ☐b ☐c ☐d

20. $\frac{3}{8}$

  $+ \frac{3}{4}$

a. $1\frac{1}{8}$ b. $\frac{8}{9}$ c. $\frac{6}{32}$ d. $\frac{3}{16}$

☐a ☐b ☐c ☐d

21. $\frac{3}{11} + \frac{1}{3} =$ a. $\frac{4}{14}$ b. $\frac{20}{33}$ c. $\frac{2}{7}$ d. $\frac{9}{11}$

☐a ☐b ☐c ☐d

22. $\frac{1}{4} + \frac{5}{7} =$ a. $\frac{7}{9}$ b. $\frac{8}{9}$ c. $\frac{7}{20}$ d. $\frac{27}{28}$

☐a ☐b ☐c ☐d

23. $86\frac{3}{16}$

  $+ 91\frac{5}{16}$

a. $177\frac{1}{2}$ b. $176\frac{8}{16}$ c. $176\frac{1}{2}$ d. $15\frac{1}{8}$

☐a ☐b ☐c ☐d

24. $24\frac{5}{16}$

  $+ 79\frac{1}{16}$

a. $103\frac{4}{16}$ b. 104   c. $103\frac{2}{8}$ d. $103\frac{3}{8}$

☐a ☐b ☐c ☐d

25. $46\frac{1}{8}$ a. $51\frac{5}{24}$ b. $51\frac{2}{20}$ c. $51\frac{1}{10}$ d. $51\frac{2}{5}$ ⓐ ⓑ ⓒ ⓓ

     $+ 5\frac{1}{12}$

26. $14\frac{1}{2}$ a. $48\frac{3}{5}$ b. $48\frac{9}{10}$ c. $49\frac{5}{7}$ d. $49\frac{2}{7}$ ⓐ ⓑ ⓒ ⓓ

     $+ 34\frac{2}{5}$

27. $\frac{10}{17} - \frac{2}{17} =$ a. $1\frac{5}{12}$ b. $\frac{12}{17}$ c. $\frac{102}{17}$ d. $\frac{8}{17}$ ⓐ ⓑ ⓒ ⓓ

28. $\frac{6}{9} - \frac{2}{9} =$ a. $\frac{4}{9}$ b. $\frac{8}{9}$ c. 4 d. $\frac{4}{18}$ ⓐ ⓑ ⓒ ⓓ

29. $\frac{3}{4} - \frac{1}{8} =$ a. $\frac{2}{4}$ b. $\frac{5}{8}$ c. $\frac{4}{12}$ d. 2 ⓐ ⓑ ⓒ ⓓ

30. $\frac{2}{3} - \frac{4}{9} =$ a. 6 b. $\frac{2}{9}$ c. $\frac{2}{6}$ d. $\frac{1}{2}$ ⓐ ⓑ ⓒ ⓓ

31. $\frac{7}{10}$ a. $\frac{8}{13}$ b. $\frac{6}{7}$ c. $\frac{11}{30}$ d. $\frac{8}{30}$ ⓐ ⓑ ⓒ ⓓ

     $- \frac{1}{3}$

32. $\frac{2}{11}$ a. $\frac{1}{2}$ b. $\frac{7}{99}$ c. $\frac{3}{20}$ d. $\frac{1}{11}$ ⓐ ⓑ ⓒ ⓓ

     $- \frac{1}{9}$

33. $\frac{99}{100}$ a. $\frac{95}{75}$ b. $\frac{83}{100}$ c. $\frac{103}{125}$ d. $\frac{103}{100}$ ⓐ ⓑ ⓒ ⓓ

     $- \frac{4}{25}$

34. $\frac{3}{4}$ a. $\frac{7}{20}$ b. $\frac{5}{9}$ c. $\frac{1}{1}$ d. $\frac{1}{9}$ ⓐ ⓑ ⓒ ⓓ

     $- \frac{2}{5}$

35. $5\frac{2}{3}$ a. $4\frac{2}{9}$ b. $5\frac{4}{9}$ c. $5\frac{8}{9}$ d. $5\frac{2}{9}$ ⓐ ⓑ ⓒ ⓓ

     $- \frac{2}{9}$

36. $14\frac{5}{12}$ a. $26\frac{6}{15}$ b. $1\frac{1}{12}$ c. $2\frac{1}{12}$ d. $2\frac{4}{9}$ ⓐ ⓑ ⓒ ⓓ

     $- 12\frac{1}{3}$

37. $7\frac{2}{5}$ a. $4\frac{1}{5}$ b. $4\frac{5}{10}$ c. $2\frac{9}{10}$ d. $4\frac{1}{10}$ ⓐ ⓑ ⓒ ⓓ

     $- 3\frac{3}{10}$

38. $376\frac{7}{33}$ a. $165\frac{1}{33}$ b. $165\frac{5}{22}$ c. $164\frac{1}{5}$ d. $164\frac{1}{33}$ ☐a ☐b ☐c ☐d

     $- 211\frac{2}{11}$

39. 74 a. $68\frac{1}{8}$ b. $67\frac{7}{8}$ c. $68\frac{7}{8}$ d. 70 ☐a ☐b ☐c ☐d

     $- 6\frac{1}{8}$

40. 29 a. $11\frac{7}{15}$ b. $47\frac{7}{15}$ c. $10\frac{8}{15}$ d. $10\frac{7}{15}$ ☐a ☐b ☐c ☐d

     $- 18\frac{7}{15}$

41. $98 - \frac{7}{9} =$ a. $98\frac{2}{9}$ b. $97\frac{7}{9}$ c. $97\frac{2}{9}$ d. $98\frac{7}{9}$ ☐a ☐b ☐c ☐d

42. $69 - 1\frac{8}{15} =$ a. $68\frac{7}{15}$ b. $67\frac{7}{15}$ c. $67\frac{6}{15}$ d. $67\frac{8}{15}$ ☐a ☐b ☐c ☐d

43. $84\frac{5}{6}$ a. $65\frac{1}{12}$ b. $75\frac{1}{3}$ c. $65\frac{1}{3}$ d. $65\frac{2}{2}$ ☐a ☐b ☐c ☐d

     $- 19\frac{3}{4}$

44. $23\frac{4}{9}$ a. $16\frac{1}{3}$ b. $6\frac{2}{27}$ c. $16\frac{7}{9}$ d. $15\frac{7}{9}$ ☐a ☐b ☐c ☐d

     $- 7\frac{2}{3}$

45. $9\frac{1}{3}$ a. $6\frac{2}{3}$ b. $6\frac{4}{9}$ c. $5\frac{2}{3}$ d. $5\frac{1}{2}$ ☐a ☐b ☐c ☐d

     $- 3\frac{5}{6}$

46. $107\frac{3}{8}$ a. $41\frac{1}{8}$ b. $40\frac{3}{16}$ c. $41\frac{1}{16}$ d. $40\frac{1}{16}$ ☐a ☐b ☐c ☐d

     $- 66\frac{5}{16}$

47. $\frac{1}{6} \times \frac{2}{9} =$ a. $\frac{2}{14}$ b. $\frac{1}{27}$ c. $\frac{3}{14}$ d. $\frac{1}{54}$ ☐a ☐b ☐c ☐d

48. $\frac{1}{33} \times \frac{11}{12} =$ a. $\frac{12}{33}$ b. $\frac{12}{45}$ c. $\frac{1}{36}$ d. $\frac{1}{45}$ ☐a ☐b ☐c ☐d

49. $16 \times \frac{3}{5} =$ a. $\frac{15}{16}$ b. $16\frac{3}{5}$ c. $9\frac{3}{5}$ d. $\frac{16}{15}$ ☐a ☐b ☐c ☐d

50. $\frac{5}{8} \times 72 =$ a. 47 b. 45 c. $\frac{7}{8}$ d. $\frac{3}{8}$ ☐a ☐b ☐c ☐d

51. $713 \times \frac{5}{9} =$ a. $\frac{59}{713}$ b. $\frac{713}{9}$ c. $396\frac{1}{9}$ d. 398 ☐a ☐b ☐c ☐d

52. $\frac{2}{3} \times 841 =$ a. 605 b. $560\frac{2}{3}$ c. $5,384$ d. $405\frac{1}{3}$ ☐a ☐b ☐c ☐d

53. $3\frac{1}{4} \times 16 =$ a. 52   b. $3\frac{1}{4}$ c. $19\frac{1}{4}$ d. 58 $\boxed{a}$ $\boxed{b}$ $\boxed{c}$ $\boxed{d}$

54. $41 \times 5\frac{5}{6} =$ a. $239\frac{1}{6}$ b. $46\frac{5}{6}$ c. $205\frac{5}{6}$ d. $\frac{55}{41}$ $\boxed{a}$ $\boxed{b}$ $\boxed{c}$ $\boxed{d}$

55. $4\frac{1}{5} \times 15 =$ a. 63   b. $4\frac{1}{15}$ c. $60\frac{1}{5}$ d. 18 $\boxed{a}$ $\boxed{b}$ $\boxed{c}$ $\boxed{d}$

56. $21 \times 6\frac{1}{3} =$ a. $126\frac{1}{3}$ b. $216\frac{1}{3}$ c. 133   d. 62 $\boxed{a}$ $\boxed{b}$ $\boxed{c}$ $\boxed{d}$

57. $4\frac{6}{7} \times 2\frac{3}{4} =$ a. $8\frac{63}{28}$ b. $12\frac{19}{14}$ c. 13   d. $13\frac{5}{14}$ $\boxed{a}$ $\boxed{b}$ $\boxed{c}$ $\boxed{d}$

58. $2\frac{2}{5} \times 12\frac{5}{6} =$ a. $31\frac{1}{5}$ b. $24\frac{7}{30}$ c. $30\frac{4}{5}$ d. 24 $\boxed{a}$ $\boxed{b}$ $\boxed{c}$ $\boxed{d}$

59. $\frac{1}{4} \div \frac{4}{7} =$ a. $\frac{7}{16}$ b. $\frac{16}{7}$ c. $\frac{4}{28}$ d. $\frac{2}{7}$ $\boxed{a}$ $\boxed{b}$ $\boxed{c}$ $\boxed{d}$

60. $\frac{1}{8} \div \frac{5}{7} =$ a. $\frac{40}{7}$ b. $\frac{5}{56}$ c. $\frac{7}{40}$ d. $11\frac{1}{5}$ $\boxed{a}$ $\boxed{b}$ $\boxed{c}$ $\boxed{d}$

61. $\frac{3}{11} \div \frac{2}{5} =$ a. $\frac{5}{16}$ b. $\frac{6}{55}$ c. $\frac{15}{22}$ d. $\frac{1}{6}$ $\boxed{a}$ $\boxed{b}$ $\boxed{c}$ $\boxed{d}$

62. $\frac{7}{8} \div \frac{2}{7} =$ a. $\frac{16}{49}$ b. $3\frac{1}{16}$ c. $\frac{5}{1}$ d. $\frac{1}{4}$ $\boxed{a}$ $\boxed{b}$ $\boxed{c}$ $\boxed{d}$

63. $\frac{1}{8} \div \frac{1}{4} =$ a. $\frac{1}{32}$ b. $\frac{4}{32}$ c. $\frac{3}{8}$ d. $\frac{1}{2}$ $\boxed{a}$ $\boxed{b}$ $\boxed{c}$ $\boxed{d}$

64. $\frac{3}{8} \div \frac{1}{16} =$ a. 6   b. $\frac{3}{128}$ c. $\frac{2}{8}$ d. 4 $\boxed{a}$ $\boxed{b}$ $\boxed{c}$ $\boxed{d}$

65. $\frac{1}{3} \div \frac{4}{9} =$ a. $\frac{4}{4}$ b. $\frac{3}{4}$ c. $\frac{4}{27}$ d. 7 $\boxed{a}$ $\boxed{b}$ $\boxed{c}$ $\boxed{d}$

66. $\frac{5}{6} \div \frac{2}{3} =$ a. $\frac{12}{15}$ b. 1   c. $\frac{5}{9}$ d. $1\frac{1}{4}$ $\boxed{a}$ $\boxed{b}$ $\boxed{c}$ $\boxed{d}$

67. $7 \div \frac{1}{5} =$ a. $6\frac{4}{5}$ b. $\frac{1}{35}$ c. $7\frac{1}{5}$ d. 35 $\boxed{a}$ $\boxed{b}$ $\boxed{c}$ $\boxed{d}$

68. $27 \div \frac{4}{7} =$ a. $47\frac{1}{4}$ b. $27\frac{4}{7}$ c. $\frac{23}{7}$ d. $3\frac{2}{7}$ $\boxed{a}$ $\boxed{b}$ $\boxed{c}$ $\boxed{d}$

69. $49 \div \frac{2}{3} =$ a. $32\frac{2}{3}$ b. $73\frac{1}{2}$ c. $\frac{47}{3}$ d. $49\frac{2}{3}$ $\boxed{a}$ $\boxed{b}$ $\boxed{c}$ $\boxed{d}$

70. $12 \div \frac{3}{8} =$ a. $4\frac{4}{9}$ b. 32   c. $12\frac{3}{8}$ d. 36 $\boxed{a}$ $\boxed{b}$ $\boxed{c}$ $\boxed{d}$

71. $\frac{1}{7} \div 24 =$ a. 24   b. $3\frac{1}{8}$ c. $\frac{1}{168}$ d. $24\frac{1}{7}$ $\boxed{a}$ $\boxed{b}$ $\boxed{c}$ $\boxed{d}$

72. $\frac{3}{5} \div 15 =$ a. 1   b. 9   c. $15\frac{3}{5}$ d. $\frac{1}{25}$ $\boxed{a}$ $\boxed{b}$ $\boxed{c}$ $\boxed{d}$

73. $\frac{1}{9} \div 11 =$ a. $\frac{1}{99}$ b. $1\frac{2}{9}$ c. $\frac{10}{11}$ d. $\frac{8}{11}$ $\boxed{a}$ $\boxed{b}$ $\boxed{c}$ $\boxed{d}$

74. $\frac{4}{5} \div 36 =$ a. 144   b. $\frac{4}{5}$ c. $\frac{1}{45}$ d. $\frac{1}{536}$ a b c d

75. $1\frac{2}{5} \div 2\frac{1}{4} =$ a. $2\frac{13}{20}$ b. $\frac{28}{45}$ c. $1\frac{1}{5}$ d. $\frac{32}{54}$ a b c d

76. $5\frac{1}{7} \div 4\frac{5}{7} =$ a. 9   b. $1\frac{4}{7}$ c. $20\frac{1}{7}$ d. $1\frac{1}{11}$ a b c d

77. $22\frac{1}{3} \div 7\frac{1}{6} =$ a. $29\frac{1}{2}$ b. 3   c. $15\frac{1}{6}$ d. $3\frac{5}{43}$ a b c d

78. $9\frac{4}{9} \div 20\frac{2}{3} =$ a. $2\frac{1}{3}$ b. 29   c. $5\frac{85}{186}$ d. $\frac{2}{3}$ a b c d

79. Mark the roman numerals for *55*.
    a. XLV   b. LV   c. XXXXXV   d. LIIIII   a b c d

80. How is this date written in arabic numerals: MCMLXX?
    a. 1952   b. 1970   c. 3120   d. 1,090,052   a b c d

81. Mrs. Cantu bought 400 yards of carpet at $16 a yard. A week later the carpet was advertised at a sale price of $13 a yard. How much could Mrs. Cantu have saved by buying the carpet during the sale?
    a. $1,200   b. $400   c. $3   d. $1,600   a b c d

82. At the supermarket Rita bought $2\frac{3}{10}$ pounds of green beans, $2\frac{1}{4}$ pounds of peas, $1\frac{7}{10}$ pounds of lima beans, $1\frac{1}{2}$ pounds of ground beef, $3\frac{1}{10}$ pounds of broccoli, and $2\frac{3}{8}$ pounds of cheese. How many pounds of vegetables did she buy?
    a. $9\frac{7}{20}$ b. $10\frac{1}{2}$ c. $10\frac{17}{20}$ d. 11   a b c d

83. Mr. Grandy buys special butter flavoring in 64-ounce jugs and repackages it in bottles holding $1\frac{1}{4}$ ounces. How many of the small bottles can be filled from a case of 6 jugs?
    a. $65\frac{1}{4}$ b. $307\frac{1}{5}$ c. $62\frac{3}{4}$ d. 1,000   a b c d

84. Velia is making items for a bake sale. For the four recipes she has selected, she needs the following amounts of flour: $2\frac{3}{4}$ cups, $1\frac{3}{4}$ cups, $2\frac{1}{2}$ cups, and $1\frac{1}{2}$ cups. How much flour does she need in all?
    a. $5\frac{7}{10}$ b. $6\frac{7}{12}$ c. $8\frac{1}{2}$ d. 1   a b c d

85. Johnson worked these hours during one pay period: $40\frac{1}{2}$ hours, 42 hours, $44\frac{1}{4}$ hours, and $43\frac{3}{4}$ hours. During the next pay period he worked 40 hours, $38\frac{3}{4}$ hours, $45\frac{1}{4}$ hours, and $42\frac{1}{2}$ hours. Find the difference between the totals of the two pay periods.
    a. 40 hours   b. $\frac{3}{4}$ hour   c. $1\frac{1}{2}$ hour   d. 4 hours   a b c d

86. The costume manager of the local theatre needs 12 masks for the group's next production. Each mask requires $\frac{3}{16}$ of a yard of fabric, but the manager has located only $1\frac{1}{2}$ yards of the type needed. How much more of the fabric must be found?

a. 1 yard   b. $\frac{3}{4}$ yard   c. 16 yards   d. $1\frac{1}{2}$ yards

a b c d

87. Ted has a parttime job. He worked $3\frac{1}{2}$ hours on Monday, $2\frac{3}{4}$ on Tuesday, and $5\frac{1}{4}$ hours on Friday. How many hours did he work last week?

a. $10\frac{1}{2}$ b. $11\frac{1}{2}$ c. $10\frac{5}{10}$ d. 10

a b c d

88. Due to limited space, the Olivers buy only $\frac{1}{2}$ of a cord of firewood at a time. Last winter they bought wood seven times. How much firewood did they buy?

a. $1\frac{3}{4}$ cords   b. $7\frac{1}{2}$ cords   c. $3\frac{1}{2}$ cords   d. 7 cords

a b c d

89. It took Mai's car $18\frac{4}{10}$ gallons of gasoline to go from Mai's home to her brother's house $423\frac{1}{5}$ miles away. How many miles per gallon did the car get?

a. 23   b. 80   c. 21   d. 18

a b c d

90. The Wilbourn triplets weighed $5\frac{3}{16}$ pounds, $4\frac{15}{16}$ pounds, and $5\frac{7}{16}$ pounds. Find the difference between the weight of the largest baby and the smallest baby.

a. $\frac{1}{2}$ pound   b. $\frac{1}{4}$ pound   c. $\frac{1}{24}$ pound   d. 1 pound

a b c d

91. A performer pays his agent $\frac{1}{10}$ of the fees he is paid. If the performer receives \$150,000 a month for a $1\frac{1}{2}$ month engagement, how much will the agent be paid?

a. \$15,000   b. \$150,000   c. \$225,000   d. \$22,500

a b c d

92. Jill barbecued nine times last month, using $\frac{1}{6}$ of a 10-pound bag of charcoal briquets each time. How much charcoal did she use?

a. 6 pounds   b. 9 pounds   c. $9\frac{1}{6}$ pounds   d. 15 pounds

a b c d

93. Sam received 3,560 patio tiles in his last order. The patios he builds take 445 tiles each. How many patios can he construct with the tiles?

a. 18   b. 80   c. 445   d. 8

a b c d

94. The water tank on a ranch holds 4,800 gallons of water. The tank is $\frac{1}{3}$ full. A windmill pumps water into the tank at a rate of 20 gallons per hour. If the wind blows continuously, how long will it take to fill the tank?

a. 160 hours   b. 12 days.   c. 16 hours   d. $\frac{1}{2}$ day

a b c d

95. A family plans to build a wooden fence around its yard. The yard has two sides thirty feet long and one side fifty feet long. Each board on the fence is $\frac{1}{3}$ foot wide When the boards are placed side-by-side, how many boards will be required to build the fence?

a. $33\frac{1}{3}$　b. 110　c. 330　d. $110\frac{1}{3}$

a b c d

96. A train has an engine 40 feet long, 92 cars 30 feet long, and a caboose 25 feet long. How long is the train?

a. 187 feet　b. 3,000 feet　c. 2,830 feet　d. 2,825 feet

a b c d

97. A box of instant potato flakes contains 16 ounces. The box will make 24 servings. How many ounces of flakes are needed for each serving?

a. $\frac{4}{6}$　b. $\frac{2}{5}$　c. $\frac{2}{3}$　d. $1\frac{1}{2}$

a b c d

98. The family car's gas tank holds 20 gallons. On a 1,200 mile trip, if the car gets 18 miles per gallon, how many times will the family need to buy gasoline?

a. 5　b. 4　c. 3　d. 2

a b c d

99. Jerry Thomas bought a new video cassette recorder. Each cassette will record for a total of $2\frac{1}{2}$ hours. Jerry wanted to record a popular television show that had 8 two-hour segments. He figured he could stop the machine for 15 minutes of commercials during each segment. How many cassettes did he need?

a. 6　b. 8　c. 4　d. 7

a b c d

100. An auto raceway has a track that is $2\frac{1}{3}$ miles around. How many laps around the track will equal 500 miles?

a. $502\frac{1}{3}$　b. $210\frac{1}{3}$　c. $214\frac{2}{7}$　d. 649

a b c d

Book 1: Fractions
Answer Key

Pages 1–2. **1.** thousands; **2.** ten millions; **3.** hundred thousands; **4.** ten; **5.** 3,811; **6.** 1,174,150; **7.** 67,015; **8.** 105,617,010; **9.** 4,233; **10.** 81,928; **11.** 353; **12.** 9,265; **13.** 1,785,696; **14.** 1,791; **15.** 70,897,803; **16.** 224,226; **17.** 397; **18.** 2,051R40; **19.** 275R9; **20.** 516R121; **21.** b; **22.** a; **23.** c; **24.** a; **25.** d; **26.** a

Page 3. **A.** **1.** ten thousands; **2.** ones; **3.** millions; **4.** hundred thousands; **5.** tens; **6.** tens; **7.** thousands; **8.** millions; **9.** ones; **10.** tens. **B.** **1.** tens; **2.** hundreds; **3.** hundreds; **4.** ones; **5.** ones; **6.** tens

Page 4. **A.** **1.** 11; **2.** 9; **3.** 12; **4.** 89; **5.** 587; **6.** 187; **7.** 139; **8.** 1,456; **9.** 15,741; **10.** 8,093; **11.** 3,206; **12.** 2,812; **13.** 129,383; **14.** 101,843; **15.** 75,013; **16.** 5,272; **17.** 63,705; **18.** 13,542; **19.** 433,061;

Page 5. **20.** 5,405; **21.** 374,470; **22.** 167,510; **23.** 77,320; **24.** 364,497. **B.** **1.** 18; **2.** 58; **3.** 87; **4.** 92; **5.** 5,069; **6.** 1,597; **7.** 2,200; **8.** 21,503; **9.** 36; **10.** 43; **11.** 45

Page 6. **A.** **1.** 1; **2.** 4; **3.** 8; **4.** 9; **5.** 137; **6.** 44; **7.** 480; **8.** 5,568; **9.** 1,580; **10.** 575; **11.** 1,776; **12.** 2,085; **13.** 207; **14.** 1,648; **15.** 162; **16.** 689; **17.** 77,638; **18.** 331. **B.** **1.** 6; **2.** 7; **3.** 1; **4.** 362; **5.** 287; **6.** 4,607; **7.** 236; **8.** 46,000; **9.** 7,250; **10.** 1,281; **11.** 4,856

Page 7. **1.** 140; **2.** 50; **3.** 2,200

Page 8. **A.** **1.** 28; **2.** 72; **3.** 56; **4.** 54; **5.** 497; **6.** 693; **7.** 348; **8.** 8,155; **9.** 1,200; **10.** 4,053,561; **11.** 486; **12.** 2,976; **13.** 1,623. **B.** **1.** 63; **2.** 81; **3.** 48; **4.** 45; **5.** 552; **6.** 729; **7.** 576; **8.** 29,997

Page 9. **A.** **1.** 4; **2.** 10; **3.** 5; **4.** 92; **5.** 82 r 4; **6.** 62; **7.** 205; **8.** 96 r 1; **9.** 5,706 r 2;

Page 10. **10.** 20; **11.** 21 r 2; **12.** 24; **13.** 1,633 r 1; **14.** 200 r 10; **15.** 8,802 r 3; **16.** 13; **17.** 393 r 3; **18.** 2,004. **B.** **1.** 6; **2.** 5; **3.** 107; **4.** 168; **5.** 340 r 1; **6.** 339; **7.** 233; **8.** 28; **9.** 568; **10.** 795 r 3; **11.** 258 r 30; **12.** 6; **13.** 63,841 miles

Page 13. John T. Brown, 3700 Millway, Hometown, MD 01234. Social Security number: 516-04-1492. **1.** 19,450.00; **2.** 125.00; **3.** 19,575.00; **4.** leave blank; **5.** 19,575.00; **6.** 1,000.00; **7.** 18,575.00; **8.** 3,450.00; **9.** 3,246.00; **10.** 214.00; **11.** leave blank

Page 14. **1.** V; **2.** III; **3.** II; **4.** IV; **5.** 8; **6.** 6; **7.** 10; **8.** 9; **9.** XI; **10.** XXIX; **11.** XV; **12.** XVIII; **13.** 50; **14.** 61; **15.** 79; **16.** 46; **17.** MDC; **18.** XCVIII; **19.** LD; **20.** DCXIII; **21.** b; **22.** a; **23.** d; **24.** b

Page 15. **A.** **1.** II; **2.** III; **3.** I; **4.** V; **5.** IV; **6.** 3; **7.** 1; **8.** 5; **9.** 2; **10.** 4; **11.** III; **12.** II. **B.** **1.** IV; **2.** I; **3.** 5; **4.** III; **5.** V; **6.** 3; **7.** II; **8.** 4; **9.** 1

Page 16. **A.** **1.** V; **2.** VIII; **3.** X; **4.** I; **5.** II; **6.** VII; **7.** III; **8.** IX; **9.** IV; **10.** VI; **11.** 8; **12.** 10; **13.** 5; **14.** 1; **15.** 2; **16.** 7; **17.** 3; **18.** 9. **B.** **1.** 8; **2.** VIII; **3.** 6; **4.** 5; **5.** I; **6.** II; **7.** 2; **8.** VII; **9.** 10; **10.** 3; **11.** IX; **12.** X; **13.** 4; **14.** VI; **15.** 3; **16.** 10; **17.** V; **18.** IV

Page 17. **A.** **1.** XI; **2.** XVIII; **3.** XIV; **4.** XVI; **5.** XII; **6.** XX; **7.** XIII; **8.** XIX; **9.** XV; **10.** XVII; **11.** XX; **12.** XXV; **13.** XXI; **14.** XXVI; **15.** XXIV; **16.** XXIX; **17.** XXII; **18.** XXVIII. **B.** **1.** XIV; **2.** XII; **3.** XVII; **4.** XXVI; **5.** XXIX; **6.** XXVIII; **7.** XXVII; **8.** XI; **9.** XVI; **10.** XX

Page 18. **A.** **1.** XXX; **2.** XXXVI; **3.** XXXI; **4.** XXXV; **5.** XXXVIII; **6.** XXXII; **7.** XXXIII; **8.** XXXVII; **9.** XXXIV; **10.** XXXIX; **11.** XL; **12.** LXX; **13.** LX; **14.** L; **15.** LXXX. **B.** **1.** XXX; **2.** L; **3.** LXX; **4.** XXXVI; **5.** XXXV; **6.** LXI; **7.** LXXVII; **8.** XL; **9.** LXVI; **10.** XXXVII

Page 19. **A.** **1.** $90 = 100 - 10$; **2.** $150 = 100 + 50$; **3.** $350 = 100 + 100 + 100 + 50$; **4.** $600 = 500 + 100$; **5.** $800 = 500 + 100 + 100 + 100$; **6.** $900 = 1,000 - 100$; **7.** $1,900 = 1,000 + 1,000 - 100$; **8.** $2,000 = 1,000 + 1,000$. **B.** **1.** MCMLXXXVI; **2.** CCIC; **3.** CCCLVI; **4.** DI; **5.** DCXIII; **6.** CML; **7.** MCCXXX; **8.** DCCLV; **9.** MM; **10.** MMI; **11.** DXVII; **12.** XCVIII; **13.** MCM; **14.** MCMXLIV; **15.** MDCCCXII

Page 20. **1.** $\frac{3}{23}$; **2.** $\frac{2}{17}$; **3.** $\frac{40}{40}$; **4.** $\frac{11}{22}$; **5.** proper fraction; **6.** improper fraction; **7.** mixed number; **8.** proper fraction; **9.** cannot be reduced; **10.** $\frac{1}{11}$; **11.** $\frac{1}{3}$; **12.** $\frac{1}{8}$; **13.** $9\frac{9}{10}$; **14.** $6\frac{2}{3}$; **15.** $12\frac{2}{3}$; **16.** 1; **17.** a; **18.** c; **19.** b

Page 21. **A.** **1.** $\frac{4}{5}$; **2.** $\frac{1}{6}$; **3.** $\frac{3}{8}$; **4.** $\frac{5}{100}$; **5.** $\frac{4}{3}$; **6.** $\frac{18}{98}$; **7.** $\frac{23}{33}$; **8.** $\frac{4}{4}$. **B.** **1.** $\frac{28}{73}$; **2.** $\frac{1}{5}$; **3.** $\frac{25}{400}$; **4.** $\frac{96}{3}$; **5.** $\frac{50}{50}$; **6.** $\frac{4}{18}$; **7.** $\frac{21}{23}$; **8.** $\frac{9}{3}$

Page 22. **A.** **1.** $\frac{16}{32}$; **2.** $\frac{1}{5}$; **3.** $\frac{4}{9}$. **B.** **1.** $\frac{1}{6}$; **2.** $\frac{7}{10}$; **3.** $\frac{2}{12}$

Page 23. **A.** **1.** improper fraction; **2.** proper fraction; **3.** mixed number; **4.** proper fraction; **5.** proper fraction; **6.** mixed number; **7.** improper fraction. **B.** **1.** improper fraction; **2.** proper fraction; **3.** proper fraction; **4.** improper fraction; **5.**

mixed number; **6.** improper fraction; **7.** mixed number

Page 24. A. 1. $\frac{1}{2}$; **2.** $\frac{2}{5}$; **3.** $\frac{1}{3}$; **4.** cannot be reduced; **5.** $\frac{3}{4}$; **6.** $\frac{1}{2}$; **7.** $\frac{1}{2}$; **8.** cannot be reduced; **9.** $\frac{1}{5}$; **10.** $\frac{1}{7}$; **11.** $\frac{3}{5}$; **12.** $\frac{1}{11}$;

Page 25. 13. $\frac{1}{4}$; **14.** cannot be reduced; **15.** cannot be reduced; **16.** $\frac{1}{3}$; **17.** $\frac{1}{5}$; **18.** cannot be reduced; **19.** $\frac{1}{4}$; **20.** $\frac{1}{4}$; **21.** $\frac{3}{4}$; **22.** $\frac{3}{10}$; **23.** cannot be reduced; **24.** $\frac{1}{4}$. **B. 1.** $\frac{1}{2}$; **2.** cannot be reduced; **3.** $\frac{1}{5}$; **4.** cannot be reduced; **5.** $\frac{1}{4}$; **6.** cannot be reduced; **7.** $\frac{1}{12}$; **8.** $\frac{1}{4}$; **9.** $\frac{3}{13}$; **10.** $\frac{1}{6}$; **11.** $\frac{1}{4}$; **12.** $\frac{4}{7}$; **13.** cannot be reduced; **14.** cannot be reduced; **15.** $\frac{6}{25}$; **16.** $\frac{2}{7}$; **17.** $\frac{1}{5}$; **18.** cannot be reduced; **19.** $\frac{8}{9}$; **20.** $\frac{1}{5}$; **21.** cannot be reduced; **22.** cannot be reduced; **23.** $\frac{1}{3}$; **24.** $\frac{1}{5}$; **25.** cannot be reduced; **26.** $\frac{6}{7}$

Page 26. A. 1. $3\frac{1}{2}$; **2.** $1\frac{1}{5}$; **3.** $3\frac{1}{4}$; **4.** $4\frac{1}{2}$; **5.** 3; **6.** $17\frac{1}{3}$; **7.** 4; **8.** $6\frac{1}{2}$; **9.** $33\frac{1}{3}$. **B. 1.** $1\frac{2}{3}$; **2.** $1\frac{1}{8}$; **3.** $10\frac{1}{2}$; **4.** $3\frac{1}{4}$; **5.** 5; **6.** $15\frac{3}{4}$; **7.** 3; **8.** $9\frac{1}{2}$; **9.** $66\frac{2}{3}$

Page 27. A. 1. 5; **2.** 15; **3.** 1; **4.** 1; **5.** 8; **6.** 39; **7.** 12; **8.** 3; **9.** 51. **B. 1.** 1; **2.** 1; **3.** 1; **4.** 1; **5.** 1; **6.** 1; **7.** 1; **8.** 1; **9.** 1; **10.** 1; **11.** 1; **12.** 1

Page 28. 2. improper, 4; **3.** proper, $\frac{2}{3}$; **4.** proper, $\frac{1}{3}$; **5.** improper, $3\frac{3}{4}$; **6.** improper, $1\frac{3}{4}$; **7.** improper, 1; **8.** proper, $\frac{4}{5}$; **9.** proper, $\frac{3}{4}$; **10.** improper, 5; **11.** improper, 1; **12.** improper, $6\frac{1}{2}$; **13.** improper, $1\frac{1}{2}$; **14.** proper, $\frac{1}{3}$; **15.** improper, 5; **16.** improper, $3\frac{1}{7}$; **17.** proper, $\frac{3}{4}$; **18.** proper, $\frac{2}{3}$; **19.** improper, $7\frac{1}{3}$; **20.** improper, $2\frac{1}{3}$; **21.** improper, $8\frac{8}{9}$; **22.** improper, 12; **23.** proper, $\frac{3}{7}$; **24.** proper, $\frac{1}{4}$

Page 29. 1. $\frac{2}{3}$; **2.** $\frac{1}{3}$; **3.** 1; **4.** 8; **5.** $1\frac{1}{3}$; **6.** $17\frac{1}{3}$; **7.** $4\frac{1}{4}$; **8.** 1; **9.** $\frac{1}{3}$; **10.** $3\frac{1}{5}$; **11.** $\frac{3}{8}$; **12.** $\frac{2}{3}$; **13.** $\frac{1}{3}$; **14.** $\frac{13}{16}$; **15.** $\frac{3}{16}$; **16.** $\frac{2}{3}$; **17.** $\frac{3}{5}$; **18.** $1\frac{2}{3}$; **19.** $2\frac{1}{2}$; **20.** $\frac{3}{14}$; **21.** $\frac{1}{7}$; **22.** $\frac{4}{5}$; **23.** $1\frac{2}{3}$; **24.** $\frac{1}{17}$; **25.** $1\frac{1}{2}$

Page 30. 2. $1,590, $190, $\frac{19}{159}$; **3.** $1,590, $450, $\frac{45}{159}$; **4.** $1,590, $44, $\frac{22}{795}$; **5.** $1,590, $100, $\frac{10}{159}$

Page 31. 1. c; **2.** b; **3.** a; **4.** c; **5.** b; **6.** a; **7.** a; **8.** d; **9.** b;

Page 32. 10. c; **11.** d; **12.** c; **13.** b; **14.** a; **15.** b; **16.** d; **17.** a; **18.** a; **19.** a; **20.** b;

Page 33. 21. b; **22.** d; **23.** a; **24.** c; **25.** c; **26.** a; **27.** d; **28.** d

Page 34. 1. $\frac{5}{8}$; **2.** $\frac{6}{7}$; **3.** $\frac{8}{15}$; **4.** $\frac{5}{9}$; **5.** $\frac{7}{8}$; **6.** $\frac{5}{6}$; **7.** $\frac{7}{10}$; **8.** $\frac{13}{14}$. **B. 1.** $\frac{11}{12}$; **2.** $\frac{4}{5}$; **3.** $\frac{10}{27}$; **4.** $\frac{11}{18}$; **5.** $\frac{9}{13}$; **6.** $\frac{20}{31}$; **7.** $\frac{11}{14}$; **8.** $\frac{20}{21}$; **9.** $\frac{2}{9}$

Page 35. 1. $\frac{2}{5}$; **2.** 1; **3.** $\frac{3}{8}$; **4.** 3; **5.** $1\frac{1}{2}$; **6.** $\frac{3}{4}$; **7.** $\frac{9}{10}$; **8.** $1\frac{1}{3}$ **B. 1.** $1\frac{2}{9}$; **2.** $\frac{4}{5}$; **3.** $1\frac{1}{4}$; **4.** 1; **5.** 2; **6.** $\frac{7}{8}$; **7.** $1\frac{1}{6}$; **8.** $1\frac{2}{7}$

Page 36. A. 1. $\frac{3}{10}$; **2.** $\frac{5}{8}$; **3.** $\frac{11}{12}$; **4.** $\frac{19}{20}$; **5.** $\frac{7}{10}$; **6.** $\frac{7}{12}$; **7.** $\frac{7}{8}$; **8.** $\frac{11}{15}$;

Page 37. 9. $\frac{7}{10}$; **10.** $\frac{11}{12}$; **11.** $\frac{13}{16}$; **12.** $\frac{4}{9}$; **13.** $\frac{13}{15}$; **14.** $\frac{11}{12}$; **15.** $\frac{7}{9}$; **16.** $\frac{7}{16}$; **17.** $\frac{7}{8}$; **18.** $\frac{7}{10}$; **19.** $\frac{9}{14}$; **20.** $\frac{9}{16}$. **B. 1.** $\frac{3}{4}$; **2.** $\frac{5}{16}$; **3.** $\frac{9}{10}$; **4.** $\frac{7}{8}$; **5.** $\frac{11}{16}$; **6.** $\frac{5}{8}$; **7.** $\frac{5}{9}$; **8.** $\frac{7}{8}$; **9.** $\frac{8}{9}$; **10.** $\frac{9}{10}$; **11.** $\frac{17}{24}$; **12.** $\frac{11}{21}$; **13.** $\frac{16}{21}$; **14.** $\frac{11}{18}$; **15.** $\frac{4}{15}$; **16.** $\frac{15}{28}$

Page 38. 1. $\frac{8}{9}$; **2.** $\frac{7}{15}$; **3.** $\frac{11}{12}$; **4.** $\frac{19}{20}$; **5.** $\frac{10}{21}$; **6.** $\frac{11}{15}$; **7.** $\frac{23}{35}$; **8.** $\frac{5}{6}$; **9.** $\frac{5}{8}$; **10.** $\frac{5}{8}$; **11.** $\frac{7}{8}$; **12.** $\frac{23}{24}$; **13.** $\frac{14}{15}$; **14.** $\frac{9}{14}$; **15.** $\frac{7}{10}$; **16.** $\frac{11}{21}$; **17.** $\frac{3}{10}$; **18.** $\frac{7}{12}$; **19.** $\frac{8}{15}$; **20.** $\frac{15}{24}$; **21.** $\frac{3}{8}$; **22.** $\frac{5}{9}$; **23.** $\frac{7}{8}$; **24.** $\frac{11}{18}$; **25.** $\frac{16}{21}$; **26.** $\frac{13}{16}$; **27.** $\frac{7}{8}$; **28.** $\frac{23}{32}$

Page 39. A. 1. $\frac{14}{15}$; **2.** $\frac{9}{14}$; **3.** $\frac{9}{10}$; **4.** $\frac{41}{42}$; **5.** $\frac{7}{8}$ acre. **B. 1.** $\frac{11}{12}$; **2.** $\frac{15}{28}$; **3.** $\frac{35}{36}$; **4.** $\frac{28}{33}$; **5.** $\frac{3}{4}$

Page 40. 1. $\frac{11}{12}$; **2.** $\frac{19}{20}$; **3.** $\frac{25}{42}$; **4.** $\frac{61}{72}$; **5.** $\frac{29}{30}$; **6.** $\frac{11}{15}$; **7.** $\frac{13}{14}$; **8.** $\frac{17}{30}$; **9.** $\frac{28}{45}$; **10.** $\frac{17}{20}$; **11.** $\frac{19}{28}$; **12.** $\frac{7}{10}$; **13.** $\frac{13}{21}$; **14.** $\frac{11}{28}$; **15.** $\frac{32}{45}$; **16.** $\frac{5}{6}$; **17.** $\frac{9}{20}$; **18.** $\frac{26}{35}$; **19.** $\frac{37}{45}$; **20.** $\frac{31}{33}$; **21.** $\frac{31}{60}$; **22.** $\frac{29}{36}$; **23.** $\frac{10}{21}$; **24.** $\frac{45}{56}$; **25.** $\frac{23}{45}$; **26.** $\frac{73}{99}$; **27.** $\frac{41}{60}$; **28.** $\frac{19}{24}$

Page 41. A. 1. $14\frac{1}{2}$; **2.** $13\frac{1}{2}$; **3.** $36\frac{1}{3}$; **4.** $8\frac{1}{2}$; **5.** $8\frac{1}{2}$; **6.** $9\frac{3}{4}$; **7.** $41\frac{2}{3}$; **8.** $19\frac{1}{2}$. **B. 1.** $11\frac{1}{2}$; **2.** $3\frac{3}{5}$; **3.** $6\frac{3}{11}$; **4.** $4\frac{2}{5}$; **5.** $8\frac{1}{2}$; **6.** $212\frac{5}{8}$; **7.** $9\frac{2}{3}$; **8.** $98\frac{1}{2}$; **9.** $1\frac{1}{3}$; **10.** $191\frac{5}{11}$

Page 42. 1. $2\frac{5}{8}$; **2.** $5\frac{1}{2}$; **3.** $7\frac{11}{15}$; **4.** $12\frac{7}{12}$; **5.** $9\frac{19}{20}$; **6.** $12\frac{11}{15}$; **7.** $11\frac{11}{12}$; **8.** $7\frac{5}{14}$. **B. 1.** $23\frac{32}{55}$; **2.** $10\frac{13}{24}$; **3.** $125\frac{2}{3}$; **4.** $11\frac{13}{15}$; **5.** $27\frac{23}{30}$; **6.** $13\frac{17}{24}$; **7.** $439\frac{5}{6}$; **8.** $6\frac{3}{4}$

Page 43. A. 1. $22\frac{1}{5}$; **2.** $16\frac{3}{8}$; **3.** $6\frac{5}{12}$; **4.** 6; **5.** $8\frac{5}{28}$; **6.** $10\frac{7}{15}$; **7.** $9\frac{7}{15}$; **8.** $10\frac{1}{6}$. **B. 1.** 5; **2.** $10\frac{1}{6}$; **3.** 10; **4.** $28\frac{5}{8}$

Page 44. 1. $\frac{3}{10}$; **2.** $\frac{2}{3}$; **3.** $2\frac{1}{4}$ tons; **4.** $3\frac{1}{3}$ cups; **5.** $10\frac{1}{4}$ pounds; **6.** $56\frac{9}{10}$ gallons; **7.** $5\frac{3}{4}$ hours; **8.** $6\frac{1}{2}$ feet; **9.** $6\frac{3}{10}$ pounds

Page 45. 1. $\frac{3}{4}$; **2.** 1; **3.** $\frac{5}{6}$; **4.** $\frac{8}{9}$; **5.** $25\frac{3}{8}$; **6.** $21\frac{1}{4}$; **7.** $30\frac{19}{30}$; **8.** $14\frac{1}{6}$; **9.** $17\frac{7}{10}$; **10.** $22\frac{1}{5}$; **11.** 1; **12.** $\frac{3}{4}$; **13.** $\frac{7}{12}$; **14.** $11\frac{1}{8}$; **15.** $7\frac{5}{8}$; **16.** $5\frac{4}{9}$; **17.**

$31\frac{5}{12}$; **18.** $20\frac{1}{10}$; **19.** $125\frac{1}{10}$; **20.** $20\frac{3}{16}$; **21.** $\frac{1}{8}$; **22.** $\frac{1}{4}$; **23.** $\frac{3}{8}$

Page 46. **1.** 1; **2.** $\frac{5}{8}$; **3.** $\frac{1}{2}$; **4.** $1\frac{8}{35}$; **5.** $10\frac{5}{12}$; **6.** $3\frac{1}{10}$; **7.** $27\frac{11}{12}$; **8.** $18\frac{1}{5}$; **9.** $9\frac{1}{3}$; **10.** $10\frac{1}{2}$; **11.** $14\frac{4}{21}$; **12.** $9\frac{1}{15}$; **13.** $5\frac{3}{8}$; **14.** $5\frac{4}{5}$; **15.** $12\frac{1}{8}$; **16.** $16\frac{8}{21}$; **17.** $12\frac{16}{21}$; **18.** $7\frac{1}{20}$; **19.** $11\frac{67}{72}$; **20.** $15\frac{7}{12}$; **21.** $\frac{3}{4}$ mile; **22.** 4; **23.** $5\frac{3}{4}$ miles

Page 48. **1. a.** 6 years, **b.** no; **2. a.** 7 years, **b.** yes ; **3. a.** $7\frac{1}{2}$ years, **b.** no; **4. a.** $8\frac{1}{4}$ years, **b.** yes

Page 49. **1.** c; **2.** a; **3.** d; **4.** c; **5.** a; **6.** b; **7.** a; **8.** d; **9.** d; **10.** b;

Page 50. **11.** c; **12.** a; **13.** b; **14.** a; **15.** d; **16.** d; **17.** a; **18.** a; **19.** c; **20.** b; **21.** c;

Page 51. **22.** d; **23.** a; **24.** c; **25.** b; **26.** a; **27.** d; **28.** c; **29.** a

Page 52. **A.** **1.** $\frac{4}{9}$; **2.** $\frac{7}{11}$; **3.** $\frac{3}{5}$; **4.** $\frac{1}{3}$; **5.** $\frac{4}{7}$; **6.** $\frac{5}{8}$; **7.** $\frac{6}{13}$; **8.** $\frac{2}{5}$. **B.** **1.** $\frac{5}{9}$; **2.** $\frac{2}{7}$; **3.** $\frac{3}{16}$; **4.** $\frac{4}{9}$; **5.** $\frac{1}{4}$; **6.** $\frac{8}{17}$; **7.** $\frac{5}{9}$; **8.** $\frac{1}{6}$

Page 53. **A.** **1.** $\frac{4}{9}$; **2.** $\frac{4}{15}$; **3.** $\frac{9}{22}$; **4.** $\frac{5}{18}$; **5.** $\frac{3}{25}$; **6.** $\frac{5}{8}$; **7.** $\frac{1}{8}$; **8.** $\frac{2}{9}$. **B.** **1.** $\frac{1}{4}$; **2.** $\frac{1}{6}$; **3.** $\frac{1}{8}$; **4.** $\frac{7}{16}$; **5.** $\frac{1}{8}$; **6.** $\frac{11}{16}$; **7.** $\frac{5}{14}$; **8.** $\frac{7}{10}$

Page 54. **A.** **1.** $\frac{1}{12}$; **2.** $\frac{4}{15}$; **3.** $\frac{5}{28}$; **4.** $\frac{1}{24}$; **5.** $\frac{1}{40}$; **6.** $\frac{7}{15}$; **7.** $\frac{5}{18}$; **8.** $\frac{11}{21}$. **B.** **1.** $\frac{17}{40}$; **2.** $\frac{1}{9}$; **3.** $\frac{4}{55}$; **4.** $\frac{1}{14}$; **5.** $\frac{1}{12}$ cup; **6.** $\frac{1}{4}$

Page 55. **A.** **1.** $\frac{1}{12}$; **2.** $\frac{1}{3}$; **3.** $\frac{1}{5}$; **4.** $\frac{1}{12}$; **5.** $\frac{3}{10}$; **6.** $\frac{7}{15}$; **7.** $\frac{1}{3}$; **8.** $\frac{1}{2}$; **9.** $\frac{5}{8}$; **10.** $\frac{4}{9}$; **11.** $\frac{1}{4}$. **B.** **1.** $\frac{1}{4}$; **2.** $\frac{1}{5}$; **3.** $\frac{1}{5}$; **4.** $\frac{1}{4}$; **5.** $\frac{1}{4}$; **6.** $\frac{1}{8}$; **7.** $\frac{1}{3}$; **8.** $\frac{1}{8}$; **9.** $\frac{3}{16}$; **10.** $\frac{1}{2}$

Page 56. **A.** **1.** $15\frac{1}{4}$; **2.** $7\frac{3}{8}$; **3.** $8\frac{1}{20}$; **4.** $5\frac{1}{4}$; **5.** $3\frac{3}{10}$; **6.** $703\frac{7}{15}$; **7.** $13\frac{4}{15}$; **8.** $4\frac{2}{21}$. **B.** **1.** $7\frac{1}{4}$; **2.** $4\frac{1}{8}$; **3.** $9\frac{5}{8}$; **4.** $112\frac{1}{6}$; **5.** $51\frac{1}{8}$; **6.** $82\frac{1}{9}$; **7.** $56\frac{3}{8}$; **8.** $11\frac{17}{40}$

Page 57. **A.** **1.** $8\frac{1}{5}$; **2.** $9\frac{1}{5}$; **3.** $9\frac{1}{4}$; **4.** $61\frac{1}{3}$; **5.** $6\frac{1}{5}$; **6.** $5\frac{5}{20}$; **7.** $3\frac{1}{2}$; **8.** $8\frac{1}{3}$. **B.** **1.** $18\frac{1}{4}$; **2.** $4\frac{4}{15}$; **3.** $6\frac{7}{12}$; **4.** $12\frac{3}{10}$; **5.** $1\frac{1}{3}$; **6.** $8\frac{1}{6}$; **7.** 23; **8.** $2\frac{1}{8}$

Page 58. **A.** **1.** $7\frac{1}{3}$; **2.** $1\frac{3}{5}$; **3.** $3\frac{5}{9}$; **4.** $12\frac{4}{5}$; **5.** $\frac{2}{3}$; **6.** $7\frac{1}{3}$; **7.** $22\frac{1}{7}$; **8.** $1\frac{1}{2}$; **9.** $2\frac{5}{8}$ yards. **B.** **1.** $84\frac{8}{9}$; **2.** $\frac{7}{10}$; **3.** $36\frac{2}{3}$; **4.** $13\frac{17}{20}$; **5.** $50\frac{4}{7}$; **6.** $26\frac{1}{2}$; **7.** $514\frac{1}{10}$

Page 59. **A.** **1.** $3\frac{2}{3}$; **2.** $1\frac{3}{5}$; **3.** $3\frac{6}{7}$; **4.** $1\frac{7}{9}$;

Page 60. **5.** $\frac{9}{11}$; **6.** $4\frac{7}{11}$; **7.** $9\frac{13}{15}$; **8.** $2\frac{1}{2}$; **9.** $5\frac{4}{5}$; **10.** $10\frac{15}{19}$; **11.** $33\frac{4}{5}$; **12.** $\frac{1}{2}$ pound, $1\frac{1}{4}$ pound. **B.** **1.** $9\frac{14}{15}$; **2.** $1\frac{2}{3}$; **3.** $1\frac{17}{19}$; **4.** $35\frac{3}{5}$; **5.** $2\frac{10}{11}$; **6.** $\frac{4}{7}$; **7.** $7\frac{8}{13}$; **8.** $21\frac{2}{3}$; **9.** $4\frac{3}{5}$; **10.** $\frac{3}{11}$; **11.** $34\frac{7}{9}$; **12.** $101\frac{6}{7}$; **13.** $4\frac{5}{8}$ yards

Page 61. **A.** **1.** $1\frac{7}{8}$; **2.** $2\frac{15}{16}$; **3.** $2\frac{5}{8}$; **4.** $12\frac{7}{8}$; **5.** $3\frac{8}{9}$; **6.** $4\frac{5}{8}$. **B.** **1.** $2\frac{1}{10}$; **2.** $1\frac{7}{8}$; **3.** $2\frac{7}{8}$; **4.** $18\frac{11}{16}$; **5.** $4\frac{1}{3}$; **6.** $6\frac{16}{21}$

Page 62. **1.** $\frac{1}{4}$ gallon; **2.** $16\frac{1}{4}$ hours; **3.** $8\frac{1}{6}$ feet; **4.** $\frac{2}{3}$ ton; **5.** $\frac{3}{13}$ yard; **6.** $\frac{3}{4}$ loaf; **7.** $1\frac{3}{4}$%; **8.** 5¢ per gallon; **9.** $4\frac{11}{15}$ square inches; **10.** $\frac{3}{4}$ ounce

Page 63. **1.** $\frac{2}{3}$; **2.** $\frac{5}{8}$; **3.** $7\frac{1}{2}$; **4.** $32\frac{1}{3}$; **5.** $2\frac{1}{6}$; **6.** $3\frac{1}{2}$; **7.** $2\frac{2}{3}$; **8.** $4\frac{1}{2}$; **9.** $1\frac{11}{12}$; **10.** $2\frac{5}{6}$; **11.** $6\frac{17}{28}$; **12.** $8\frac{2}{3}$; **13.** $8\frac{1}{6}$; **14.** $28\frac{6}{7}$; **15.** $15\frac{3}{10}$

Page 64. **1.** $\frac{5}{18}$; **2.** $\frac{1}{3}$; **3.** $3\frac{5}{7}$; **4.** $\frac{5}{12}$; **5.** $\frac{1}{4}$; **6.** $\frac{11}{18}$; **7.** $4\frac{1}{2}$; **8.** $3\frac{1}{10}$; **9.** $6\frac{5}{21}$; **10.** $1\frac{1}{2}$; **11.** $1\frac{2}{3}$; **12.** $\frac{1}{4}$; **13.** $10\frac{3}{4}$; **14.** $4\frac{3}{10}$; **15.** $102\frac{2}{3}$; **16.** $3\frac{11}{15}$; **17.** $6\frac{1}{2}$; **18.** $2\frac{7}{12}$; **19.** $11\frac{1}{3}$; **20.** 16; **21.** $12\frac{1}{3}$; **22.** $2\frac{1}{7}$

Page 66. **1.** $\frac{1}{4}$; **2.** $96\frac{3}{8}$; **3.** 55; **4.** $68\frac{1}{8}$; **5.** $101\frac{1}{2}$; **6.** 0; **7.** 98; **8.** $\frac{3}{8}$; **9.** $94\frac{5}{8}$; **10.** $\frac{3}{8}$; **11.** $\frac{1}{2}$; **12.** $\frac{5}{8}$; **13.** $5\frac{3}{8}$; **14.** $14\frac{1}{8}$

Page 67. **1.** a; **2.** c; **3.** a; **4.** d; **5.** b; **6.** b; **7.** d; **8.** c; **9.** a; **10.** d; **11.** b; **12.** a;

Page 68. **13.** a; **14.** c; **15.** d; **16.** b; **17.** d

Page 69. **A.** **1.** $\frac{1}{6}$; **2.** $\frac{3}{10}$; **3.** $\frac{8}{15}$; **4.** $\frac{2}{15}$; **5.** $\frac{3}{20}$; **6.** $\frac{4}{9}$; **7.** $\frac{6}{35}$; **8.** $\frac{15}{28}$; **9.** $\frac{5}{24}$; **10.** $\frac{4}{15}$; **11.** $\frac{5}{42}$; **12.** $\frac{9}{16}$; **13.** $\frac{15}{32}$; **14.** $\frac{1}{3}$; **15.** $\frac{1}{4}$; **16.** $\frac{1}{5}$; **17.** $\frac{1}{9}$. **B.** **1.** $\frac{1}{2}$; **2.** $\frac{1}{10}$; **3.** $\frac{1}{12}$; **4.** $\frac{1}{32}$; **5.** $\frac{1}{3}$; **6.** $\frac{8}{15}$; **7.** $\frac{4}{63}$; **8.** $\frac{3}{10}$; **9.** $\frac{1}{2}$; **10.** $\frac{1}{5}$; **11.** $\frac{3}{10}$; **12.** $\frac{1}{25}$

Page 70. **A.** **1.** $\frac{1}{5}$; **2.** $\frac{2}{5}$; **3.** $\frac{3}{14}$; **4.** $\frac{4}{15}$; **5.** $\frac{1}{2}$; **6.** $\frac{1}{4}$; **7.** $\frac{1}{4}$; **8.** $\frac{1}{3}$; **9.** $\frac{1}{5}$. **B.** **1.** $\frac{1}{10}$; **2.** $\frac{1}{16}$; **3.** $\frac{1}{7}$; **4.** $\frac{1}{12}$; **5.** $\frac{5}{11}$; **6.** $\frac{1}{6}$

Page 71. **A.** **1.** $\frac{1}{2} \times \frac{3}{4} = \frac{3}{8}$; **2.** $\frac{4}{5} \times \frac{3}{7} = \frac{12}{35}$; **3.** $\frac{1}{2} \times \frac{1}{4} = \frac{1}{8}$; **4.** $\frac{2}{3} \times \frac{9}{10} = \frac{3}{5}$; **5.** $\frac{7}{10} \times \frac{12}{21} = \frac{2}{5}$; **6.** $\frac{3}{8} \times \frac{5}{6} = \frac{5}{16}$. **B.** **1.** $\frac{1}{3} \times \frac{4}{5} = \frac{4}{15}$; **2.** $\frac{2}{3} \times \frac{4}{7} = \frac{8}{21}$; **3.** $\frac{1}{3} \times \frac{2}{5} = \frac{2}{15}$; **4.** $\frac{3}{4} \times \frac{4}{9} = \frac{1}{3}$

Page 72. **A.** **1.** $2\frac{2}{3}$; **2.** $4\frac{1}{2}$; **3.** $3\frac{3}{4}$; **4.** $1\frac{1}{3}$; **5.** 4; **6.** 2; **7.** 7; **8.** 3; **9.** 90; **10.** $2\frac{6}{7}$; **11.** $1\frac{7}{9}$; **12.** 5; **13.** 2; **14.** $2\frac{5}{8}$; **15.** $3\frac{3}{4}$; **16.** $2\frac{1}{3}$. **B.** **1.** 1; **2.** 1;

3. $1\frac{4}{5}$; 4. $2\frac{2}{3}$; 5. 12; 6. 9; 7. $4\frac{2}{5}$; 8. $\frac{1}{8}$; 9. $8\frac{2}{3}$; 10. 18

Page 73. A. 1. $\frac{5}{3}$; 2. $\frac{15}{4}$; 3. $\frac{26}{5}$; 4. $\frac{38}{3}$; 5. $\frac{91}{2}$; 6. $\frac{8}{5}$. B. 1. $\frac{9}{5}$; 2. $\frac{14}{3}$; 3. $\frac{37}{6}$; 4. $\frac{55}{4}$; 5. $\frac{101}{3}$; 6. $\frac{37}{4}$

Page 74. A. 1. $3\frac{1}{3}$; 2. $1\frac{1}{3}$; 3. $14\frac{2}{3}$; 4. $5\frac{1}{3}$; 5. $1\frac{1}{8}$; 6. $9\frac{1}{3}$; 7. $1\frac{4}{7}$; 8. $1\frac{1}{15}$; 9. 9; 10. $2\frac{1}{2}$; 11. $4\frac{4}{7}$; 12. $8\frac{1}{6}$; 13. $4\frac{1}{6}$; 14. $1\frac{1}{6}$; 15. $5\frac{1}{10}$. B. 1. $5\frac{2}{5}$; 2. $1\frac{1}{2}$; 3. $12\frac{3}{4}$; 4. $4\frac{1}{4}$; 5. $1\frac{5}{8}$; 6. $5\frac{1}{9}$; 7. $4\frac{3}{4}$; 8. $3\frac{1}{3}$; 9. $5\frac{1}{3}$; 10. $4\frac{1}{2}$; 11. $4\frac{1}{14}$; 12. $3\frac{4}{25}$; 13. $3\frac{1}{3}$; 14. $5\frac{6}{7}$; 15. $11\frac{2}{3}$

Page 75. A. 1. 10; 2. $1\frac{1}{2}$; 3. $7\frac{1}{3}$; 4. $1\frac{4}{5}$; 5. $16\frac{2}{3}$; 6. $12\frac{1}{3}$; 7. $6\frac{1}{4}$; 8. $20\frac{4}{5}$; 9. $24\frac{1}{2}$; 10. $9\frac{2}{7}$. B. 1. 20; 2. $1\frac{3}{4}$; 3. $16\frac{1}{4}$; 4. $\frac{2}{3}$; 5. $12\frac{1}{3}$; 6. $\frac{2}{3}$; 7. $25\frac{5}{11}$; 8. $7\frac{53}{63}$; 9. $8\frac{5}{9}$; 10. $4\frac{2}{5}$; 11. $\frac{2}{15}$; 12. $8\frac{67}{100}$

Page 76. 1. $\frac{1}{6}$ pound; 2. $\frac{1}{2}$ acre; 3. $350; 4. no; 5. $5\frac{5}{8}$ cups flour, $4\frac{3}{8}$ quarts milk, $1\frac{1}{4}$ pounds butter; 6. 4 yards; 7. 55 miles; 8. $1\frac{1}{2}$ hours

Page 77. 1. $\frac{5}{8}$; 2. $2\frac{1}{4}$; 3. 3; 4. $\frac{7}{9}$; 5. $1\frac{1}{20}$; 6. 12; 7. 10; 8. $\frac{1}{9}$; 9. $\frac{3}{8}$; 10. 1; 11. $\frac{3}{5}$; 12. $1\frac{1}{5}$; 13. 5; 14. $\frac{19}{20}$; 15. $\frac{5}{8}$; 16. $15\frac{3}{5}$; 17. 8; 18. $\frac{1}{25}$; 19. $\frac{1}{6}$; 20. 1; 21. $\frac{1}{4}$; 22. $8\frac{7}{10}$; 23. $\frac{1}{2}$; 24. 35; 25. $1\frac{1}{2}$; 26. $67\frac{5}{9}$; 27. 1; 28. $3\frac{3}{4}$; 29. $18\frac{2}{3}$; 30. 21; 31. $4\frac{1}{2}$; 32. $14\frac{2}{9}$; 33. $\frac{1}{6}$; 34. 9

Page 78. 1. 15 pounds; 2. $1\frac{1}{4}$ pounds; 3. $1\frac{1}{4}$ pounds; 4. 6 pounds; 5. 7 pounds; 6. $4\frac{1}{2}$ pounds; 7. 4 pounds

Page 79. 1. b; 2. a; 3. c; 4. b; 5. d; 6. a; 7. a; 8. b; 9. d; 10. c; 11. c; 12. d; 13. a;

Page 80. 14. b; 15. c; 16. a; 17. c; 18. d; 19. b; 20. a; 21. d; 22. d; 23. b; 24. a;

Page 81. 25. c; 26. a; 27. d; 28. b; 29. a; 30. b

Page 82. A. 1. $\frac{3}{4}$; 2. $\frac{3}{4}$; 3. $\frac{4}{9}$; 4. $\frac{3}{10}$; 5. $\frac{9}{14}$; 6. $\frac{9}{10}$; 7. $\frac{5}{6}$; 8. $4\frac{2}{3}$; 9. $\frac{6}{7}$; 10. $1\frac{1}{11}$; 11. $2\frac{3}{16}$; 12. $\frac{9}{14}$. B. 1. $\frac{8}{9}$; 2. $\frac{5}{8}$; 3. $\frac{16}{25}$; 4. $\frac{15}{32}$; 5. $\frac{2}{3}$; 6. $\frac{33}{35}$; 7. $1\frac{7}{9}$; 8. $\frac{3}{5}$; 9. $\frac{2}{7}$; 10. $\frac{5}{22}$

Page 83. A. 1. $\frac{1}{2} \div \frac{3}{4}$; 2. $\frac{3}{5} \div \frac{1}{2}$; 3. $\frac{2}{3} \div \frac{1}{4}$; 4. $\frac{2}{5} \div \frac{2}{5}$; 5. $\frac{4}{7} \div \frac{1}{8}$; 6. $\frac{7}{8} \div \frac{3}{4}$; 7. $\frac{1}{2} \div \frac{5}{6}$; 8. $\frac{1}{3} \div \frac{1}{9}$; 9. $\frac{2}{7} \div \frac{1}{4}$; 10. $\frac{1}{6} \div \frac{1}{2}$. B. 1. $\frac{2}{3} \div \frac{3}{8}$; 2. $\frac{3}{4} \div \frac{1}{2}$; 3. $\frac{7}{8} \div \frac{2}{3}$; 4. $\frac{3}{8} \div \frac{1}{4}$; 5. $\frac{4}{7}$; 6. $\frac{2}{7} \div \frac{3}{4}$; 7. $\frac{1}{4} \div \frac{2}{11}$; 8. $\frac{2}{9} \div \frac{3}{8}$; 9. $\frac{2}{5}$; 10. $\frac{2}{3} \div \frac{1}{6}$

Page 84. A. 1. $\frac{2}{3}$; 2. $\frac{1}{2}$; 3. $\frac{4}{7}$; 4. $\frac{4}{5}$; 5. $\frac{1}{4}$; 6. $\frac{1}{2}$; 7. $\frac{2}{3}$; 8. $\frac{3}{4}$;

Page 85. 9. $\frac{1}{7}$; 10. $\frac{3}{4}$; 11. $\frac{1}{5}$; 12. $\frac{1}{5}$; 13. $\frac{10}{21}$; 14. $\frac{4}{21}$. B. 1. 1; 2. 1; 3. $\frac{5}{9}$; 4. $\frac{5}{16}$; 5. $\frac{3}{4}$; 6. 1; 7. $\frac{2}{3}$; 8. $\frac{1}{2}$; 9. $\frac{3}{7}$; 10. 1; 11. $\frac{3}{10}$; 12. $\frac{1}{4}$; 13. $\frac{1}{3}$; 14. $\frac{1}{4}$

Page 86. A. 1. $1\frac{1}{15}$; 2. $1\frac{7}{8}$; 3. $1\frac{1}{8}$; 4. $\frac{3}{4}$; 5. $1\frac{3}{4}$; 6. $\frac{6}{7}$; 7. $\frac{6}{7}$; 8. $\frac{15}{44}$; 9. $1\frac{1}{4}$; 10. 4; 11. $2\frac{1}{2}$; 12. 2; 13. $1\frac{1}{2}$; 14. $2\frac{2}{5}$; 15. $1\frac{1}{3}$; 16. $1\frac{3}{4}$; 17. $1\frac{1}{4}$; 18. $2\frac{1}{3}$; 19. $5\frac{1}{4}$; 20. $1\frac{1}{2}$;

Page 87. 21. $1\frac{1}{3}$; 22. 4; 23. $2\frac{6}{7}$; 24. $1\frac{4}{21}$; 25. $1\frac{2}{7}$; 26. $1\frac{1}{4}$; 27. $8\frac{4}{5}$; 28. $1\frac{1}{8}$. B. 1. 1; 2. 1; 3. $1\frac{3}{5}$; 4. $1\frac{3}{5}$; 5. $3\frac{1}{2}$; 6. $\frac{1}{2}$; 7. $1\frac{1}{2}$; 8. $\frac{2}{3}$; 9. $3\frac{1}{3}$; 10. $2\frac{2}{3}$; 11. $1\frac{3}{5}$; 12. $2\frac{1}{4}$; 13. $1\frac{1}{4}$; 14. $3\frac{1}{5}$; 15. $4\frac{2}{7}$; 16. $2\frac{2}{9}$; 17. $11\frac{2}{3}$; 18. $1\frac{1}{2}$; 19. $2\frac{2}{11}$; 20. $6\frac{3}{10}$; 21. $2\frac{8}{11}$; 22. $2\frac{1}{3}$; 23. $2\frac{2}{3}$; 24. $\frac{3}{10}$

Page 88. A. 1. $\frac{6}{1}$; 2. $\frac{23}{1}$; 3. $\frac{18}{1}$; 4. $\frac{100}{1}$; 5. $\frac{5}{1}$; 6. $\frac{50}{1}$; 7. $\frac{10}{1}$; 8. $\frac{2}{1}$; 9. $\frac{1}{1}$; 10. $\frac{7}{1}$; 11. $\frac{101}{1}$; 12. $\frac{44}{1}$; 13. $\frac{2}{1}$; 14. $\frac{51}{1}$; 15. $\frac{27}{1}$; 16. $\frac{24}{1}$; 17. $\frac{23}{1}$; 18. $\frac{10}{1}$. B. 1. $\frac{11}{1}$; 2. $\frac{20}{1}$; 3. $\frac{45}{1}$; 4. $\frac{3}{1}$; 5. $\frac{19}{1}$; 6. $\frac{100}{1}$; 7. $\frac{21}{1}$; 8. $\frac{36}{1}$; 9. $\frac{145}{1}$; 10. $\frac{110}{1}$; 11. $\frac{135}{1}$; 12. $\frac{119}{1}$; 13. $\frac{39}{1}$; 14. $\frac{77}{1}$; 15. $\frac{4}{1}$; 16. $\frac{14}{1}$; 17. $\frac{1}{1}$; 18. $\frac{5}{1}$; 19. $\frac{38}{1}$; 20. $\frac{98}{1}$; 21. $\frac{16}{1}$; 22. $\frac{76}{1}$; 23. $\frac{40}{1}$; 24. $\frac{84}{1}$

Page 89. A. 1. $4\frac{1}{2}$; 2. $12\frac{1}{2}$; 3. 45; 4. 63; 5. 36; 6. 16; 7. 69; 8. 78; 9. 22; 10. 42. B. 1. 6; 2. $3\frac{1}{3}$; 3. $53\frac{1}{3}$; 4. 28; 5. 25; 6. 135; 7. $37\frac{1}{3}$; 8. 108; 9. 14; 10. 57

Page 90. A. 1. $\frac{2}{27}$; 2. $\frac{3}{25}$; 3. $\frac{1}{40}$; 4. $\frac{1}{77}$; 5. $\frac{1}{3}$; 6. $\frac{1}{5}$; 7. $\frac{1}{32}$; 8. $\frac{1}{7}$; 9. $\frac{2}{15}$; 10. $\frac{1}{49}$. B. 1. $\frac{1}{12}$; 2. $\frac{1}{24}$; 3. $\frac{1}{16}$; 4. $\frac{1}{35}$; 5. $\frac{1}{21}$; 6. $\frac{1}{12}$

Page 91. A. 1. $1\frac{1}{15}$; 2. $\frac{6}{11}$; 3. $1\frac{8}{27}$; 4. $3\frac{3}{4}$; 5. $\frac{25}{48}$; 6. $\frac{5}{9}$; 7. $\frac{3}{14}$; 8. $2\frac{6}{13}$; 9. $\frac{2}{13}$; 10. $\frac{1}{3}$. B. 1. $\frac{7}{8}$; 2. $1\frac{1}{3}$; 3. $\frac{48}{49}$; 4. $2\frac{1}{3}$; 5. $\frac{33}{35}$; 6. $\frac{2}{15}$

Page 92. 1. 61 (The trotline will be divided into 60 segments, but Mr. Robinson will also need a hook at the beginning of the trotline.); 2. 24 cans; 3. $\frac{11}{15}$ ton; 4. $\frac{3}{8}$ hour; 5. $1\frac{1}{4}$ hours; 6. 8 yards; 7. 20 towels; 8. $\frac{11}{20}$ pound

Page 93. 1. 2; 2. $\frac{1}{10}$; 3. $\frac{2}{3}$; 4. 2; 5. 12; 6. $6\frac{2}{3}$; 7. 25; 8. $12\frac{1}{2}$; 9. $\frac{1}{16}$; 10. $\frac{1}{3}$; 11. 21; 12. $3\frac{1}{3}$; 13. $\frac{1}{4}$; 14. $2\frac{1}{2}$; 15. 8; 16. 26; 17. $\frac{25}{36}$; 18. $\frac{16}{35}$; 19. $\frac{1}{4}$; 20. $\frac{1}{2}$; 21. $1\frac{3}{7}$; 22. $3\frac{9}{37}$; 23. $7\frac{13}{20}$; 24. $\frac{9}{70}$; 25. 21;

26. $28\frac{5}{6}$; **27.** $29\frac{1}{3}$; **28.** $48\frac{3}{8}$; **29.** $\frac{1}{32}$; **30.** $1\frac{1}{34}$

Page 94. **1.** $\frac{1}{2}$; **2.** $\frac{1}{6}$; **3.** $\frac{5}{8}$; **4.** 2; **5.** 15; **6.** $1\frac{1}{5}$; **7.** 49; **8.** 5; **9.** $\frac{1}{10}$; **10.** $\frac{2}{3}$; **11.** $1\frac{1}{2}$; **12.** $\frac{2}{5}$; **13.** $1\frac{2}{5}$; **14.** $\frac{1}{16}$; **15.** $\frac{1}{5}$; **16.** $7\frac{1}{2}$; **17.** $12\frac{1}{2}$; **18.** $\frac{1}{2}$; **19.** $\frac{1}{8}$; **20.** $5\frac{1}{3}$; **21.** 9; **22.** $2\frac{11}{20}$ pounds

Page 95. The better buy in each pair is brand X beans, king size detergent, cheese pizza, and frozen turkey dinner.

Page 96. **1.** c; **2.** a; **3.** a; **4.** a; **5.** d; **6.** d; **7.** a; **8.** b; **9.** c; **10.** c; **11.** d;

Page 97. **12.** a; **13.** b; **14.** b; **15.** a; **16.** c; **17.** d; **18.** a; **19.** c; **20.** a; **21.** b; **22.** d; **23.** a; **24.** d;

Page 98. **25.** a; **26.** b; **27.** d; **28.** a; **29.** b; **30.** b; **31.** c; **32.** b; **33.** b; **34.** a; **35.** b; **36.** c; **37.** d;

Page 99. **38.** a; **39.** b; **40.** c; **41.** c; **42.** b; **43.** a; **44.** d; **45.** d; **46.** c; **47.** b; **48.** c; **49.** c; **50.** b; **51.** c; **52.** b;

Page 100. **53.** a; **54.** a; **55.** a; **56.** c; **57.** d; **58.** c; **59.** a; **60.** c; **61.** c; **62.** b; **63.** d; **64.** a; **65.** b; **66.** d; **67.** d; **68.** a; **69.** b; **70.** b; **71.** c; **72.** d; **73.** a;

Page 101. **74.** c; **75.** b; **76.** d; **77.** d; **78.** c; **79.** b; **80.** b; **81.** a; **82.** a; **83.** b; **84.** c; **85.** d;

Page 102. **86.** b; **87.** b; **88.** c; **89.** a; **90.** a; **91.** d; **92.** d; **93.** d; **94.** a;

Page 103. **95.** c; **96.** d; **97.** c; **98.** b; **99.** a; **100.** c